Workshop on New Scientific Opportunities with the TRIUMF ARIEL e-linac

Journal of Physics: Conference Series Volume 2391

Vancouver, Canada
25 – 27 May 2022

ISBN: 978-1-7138-6496-7
ISSN: 1742-6588

Printed from e-media with permission by:

Curran Associates, Inc.
57 Morehouse Lane
Red Hook, NY 12571

CURRAN ASSOCIATES INC. proceedings.com

Some format issues inherent in the e-media version may also appear in this print version.

Institute of Physics
Dirac House, Temple Back
Bristol BS1 6BE UK

Phone: 44 1 17 929 7481
Fax: 44 1 17 920 0979

techtracking@iop.org

Additional copies of this publication are available from:

Curran Associates, Inc.
57 Morehouse Lane
Red Hook, NY 12571 USA
Phone: 845-758-0400
Fax: 845-758-2633
Email: curran@proceedings.com
Web: www.proceedings.com

TABLE OF CONTENTS

PREFACE

These are the proceedings of the "New Scientific Opportunities at the TRIUMF ARIEL e-linac" workshop, which took place at TRIUMF laboratory, in Vancouver, Canada, May 25-27, 2022. The workshop was a hybrid of in-person and online attendance via Zoom, gathering together theorists and experimentalists with shared interests in MeV-scale physics at the intensity frontier, in order to stimulate ideas and collaboration for novel applications of new, high-intensity, modest-energy accelerators like the ARIEL e-Linac.

While the LHC and its large detector collaborations continue to seek BSM physics at the energy frontier, and increasingly large subterranean detectors probe neutrino and dark matter signals with greater and greater sensitivities, the increasing scale of these experiments represents a substantial undertaking, in terms of cost and timeline. These have yielded important results, confirming the existence of the Higgs and furthering our understanding of neutrino dynamics, but they have not yet provided a clear view of what lies beyond the Standard Model. The prevailing expectations of the previous decades—minimal supersymmetry, WIMP-based dark matter—have not yet been observed.

A complementary experimental thrust in the quest for new physics is to probe lower energy scales at the intensity and precision frontiers. Such experiments can be mounted more nimbly, and in parallel, to focus on different reported anomalies as lampposts, to improve precision in places where new physics may lurk, and to test new classes of BSM physics that may arise at those energies. With no definitive guidance on the form an underlying theory must take, this is an increasingly appealing approach to the search.

Modern accelerator designs can provide higher and higher beam currents, achieving high luminosities without the need for thick targets, unlocking new experimental avenues. Energy Recovery Linacs (ERLs) at the sub-GeV scale are a particularly exciting emerging platform, with a series of machines at various levels of development, including the planned upgrade for ARIEL, the MESA accelerator complex currently under construction, and the planned PERLE facility.

The articles in these proceedings showcase the experimental avenues now being explored, or which could be undertaken at ARIEL and similar-scale accelerators, as well as theoretical studies in these regimes, and the status of accelerators that will enable these and future experiments that aim to address current anomalies and outstanding questions in particle and nuclear physics.

The workshop was funded by the Gordon and Betty Moore Foundation, to which the organizing committee give their sincere thanks.

Jan Bernauer (CFNS, Stony Brook University and RIKEN BNL Research Center)
Ross Corliss (CFNS, Stony Brook University)
Michael Hasinoff (University of British Columbia)
Rituparna Kanungo (Saint Mary's University)
Jeffery Martin (University of Winnipeg)
Richard Milner (Massachusetts Institute of Technology)
Katherine Pachal (TRIUMF)
Stanley Yen (TRIUMF)

Journal of Physics: Conference Series **2391** (2022) 011002 doi:10.1088/1742-6596/2391/1/011002

Peer Review Statement

All papers published in this volume have been reviewed through processes administered by the Editors. Reviews were conducted by expert referees to the professional and scientific standards expected of a proceedings journal published by IOP Publishing Publishing.

- **Type of peer review:** Single Anonymous
- **Conference submission management system:** Morressier
- **Number of submissions received:** 19
- **Number of submissions sent for review:** 19
- **Number of submissions accepted:** 19
- **Acceptance Rate (Submissions Accepted / Submissions Received × 100):** 100
- **Average number of reviews per paper:** 1
- **Total number of reviewers involved:** 7
- **Contact person for queries:**
  Name: Ross Corliss
  Email: ross.corliss@stonybrook.edu
  Affiliation: Center for Frontiers in Nuclear Science, Stony Brook University

Journal of Physics: Conference Series **2391** (2022) 012001 doi:10.1088/1742-6596/2391/1/012001

Scientific Opportunities at the ARIEL Electron Linac

Jan Bernauer[1,2], Ross Corliss[1], Susan Gardner[3], Michael Hasinoff[4], Rituparna Kanungo[5], Jeffery Martin[6], Richard Milner[7], Katherine Pachal[8], Toshimi Suda[9] and Stanley Yen[8]

[1] Stony Brook University, Stony Brook, NY, USA
[2] RIKEN BNL Research Center, Upton, NY, USA
[3] University of Kentucky, Lexington, KY, USA
[4] University of British Columbia, Vancouver, BC, Canada
[5] St. Mary's University, Halifax, NS, Canada
[6] University of Winnipeg, Winnipeg, MB, Canada
[7] Massachusetts Institute of Technology, Cambridge, MA, USA
[8] TRIUMF, Vancouver, BC, Canada
[9] Tohoku University, Sendai, Japan

E-mail: ross.corliss@stonybrook.edu

Abstract. This paper gives an overview of the scientific opportunities at the ARIEL electron accelerator identified in open discussion at the workshop, including applications in hadron structure, astrophysical processes, tests of quantum electrodynamics, dark matter and other BSM physics, and material science.

1. Introduction

The Advanced Rare Isotope Laboratory (ARIEL), now under construction, will extend TRIUMF's existing isotope program, enabling world-class research on the nature of atomic nuclei, the origin of the heavy chemical elements, and contributing more broadly to nuclear medicine, materials science, and nuclear and particle physics research. ARIEL is driven jointly by the TRIUMF cyclotron and by a newly constructed superconducting electron accelerator (e-linac) which will enable isotope production via photo-production and photo-fission. Designed to provide a continuous, 10 mA beam with energies up to 30 MeV (300 kW), and with several upgrade paths available (see T. Planches's contribution to these proceedings for details), this e-linac can also enable new physics programs independently of the ISAC facility, both in its current form and through those upgrades.

In the open discussion sessions of the May, 2022 workshop, a number of possible new scientific opportunities at the e-linac were identified. We have collected them here with the purpose of providing a starting point for the interested reader to consider them in more depth.

2. Hadron Structure and Astrophysics

2.1. Low Q^2 Elastic Electron Scattering

Non-perturbative Quantum Chromodynamics describes the physics inside nucleons and nuclei, with many of the bulk properties emerging from the dynamics of the system instead of being determined by the properties of the constituents—in contrast, for example, to the QED description of atoms. Elastic electron-hadron scattering allows us to directly access these intrinsic, emergent properties, validating our current theoretical model. Precision measurements are benchmarks for lattice calculations and provide important input to other fields, from astrophysics to atomic physics.

2.1.1. Proton Targets Using electron-proton scattering on a hydrogen-containing target, one can determine the elastic form factors G_E and G_M, which describe the distribution of charge and currents inside the proton. Intrinsic properties of the proton can be related to these, such as the charge and magnetic radii of the proton, which are related to the slope of these form factors at zero four-momentum transfer. Other emergent phenomena, such as a meson cloud that forms around the bare nucleon, leave signatures in the shape of the form factors as well.

To first order, the elastic electron-proton scattering cross section can be written as

$$\frac{\mathrm{d}\sigma}{\mathrm{d}\Omega} = \frac{1}{\epsilon(1+\tau)}\left[\epsilon G_E^2 + \tau G_M^2\right]\left(\frac{\mathrm{d}\sigma}{\mathrm{d}\Omega}\right)_{\text{Mott}}, \tag{1}$$

where the photon polarization ϵ varies between 0 and 1 for backward to forward scattering, and $\tau = Q^2/(4m^2_{\text{proton}})$ is proportional to the exchanged four-momentum squared. The electric and magnetic radii are given by

$$\left\langle r_{E(M)}\right\rangle^2 = -\frac{6\hbar^2}{G_{E(M)}(0)}\left.\frac{\mathrm{d}G_{E(M)}}{\mathrm{d}Q^2}\right|_{Q^2=0}, \tag{2}$$

in both cases requiring an extrapolation of the measured form factor to $Q^2 = 0$.

The charge radius also plays a role in spectroscopy as part of the Lamb-shift. It has attracted interest over the last decade, in form of the so-called proton radius puzzle, originally established in 2010 by a 4 % difference between extractions using muonic spectroscopy [1] ($r_\text{p} = 0.841\,84(67)$ fm) and both the results of the Mainz high-precision form factor experiment [2] ($r_\text{p} = 0.879(5)_\text{stat}(6)_\text{syst}$ fm) and the CODATA value [3] ($r_\text{p} = 0.8768(69)$ fm), based on a series of normal hydrogen spectroscopy measurements and radius extractions from earlier scattering data.

Recent measurements, especially in spectroscopy, tend to agree with the smaller radius, however some measurements prefer the larger. In the regime of scattering, the recent PRad [4] result gives the smaller result but exhibits a clear difference from all earlier measurements not only in the extrapolation to $Q^2 = 0$, but also over the whole overlap with earlier measurements (see Fig. 1), putting our knowledge about the proton form factors, not only at small Q^2, into question.

To minimize the effect of the extrapolation, experiments try to reach smaller and smaller Q^2. PRad does so by instrumenting extremely small scattering angles at comparatively large beam energies. A complementary measurement would measure the form factors in the same Q^2 range by instrumenting larger scattering angles using smaller beam energies. This would also allow the measurement of backward angles and the extraction of the magnetic radius; since the contribution of G_M is suppressed by $\tau \propto Q^2$, data is sensitive to G_M only if ε is small, i.e., if θ is close to 180°. Currently, data at low Q^2 are measured at higher beam energies, moving data sensitive to G_M to larger Q^2. Therefore any extraction relies on the stiffness of the fit to extrapolate from larger Q^2. The Mainz fit [2], which is relatively flexible, finds some structure at $Q^2 = 0.03\,(\text{GeV}/c)^2$, while other fits with less flexibility do not resolve this structure. This leads to large variations in the extracted value for the magnetic radius. Similar to the charge radius, the low-Q^2 shape of the magnetic form factor is important for spectroscopy via the so-called Zemach-radius. A reliable determination of the magnetic radius and low-Q^2 magnetic

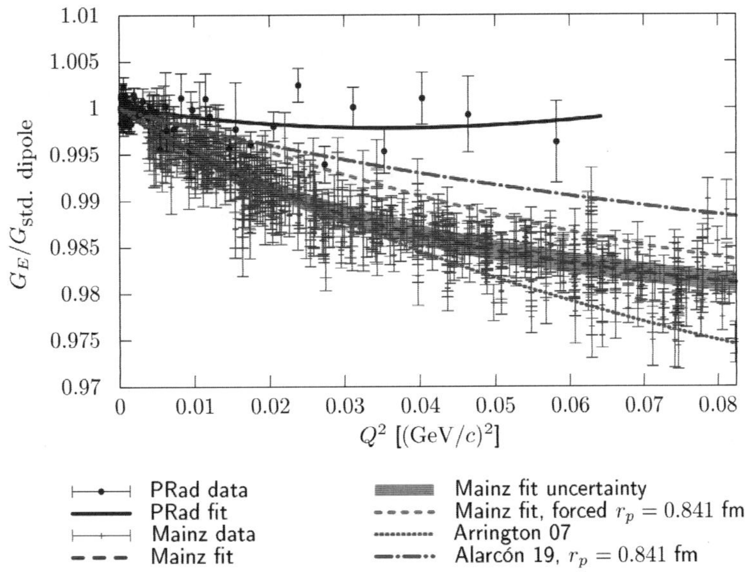

Figure 1. Recent results for G_E, normalized to the standard dipole. Red data points are the Mainz data [2] with fit and error band. Additionally, a fit to the Mainz data with a radius forced to the small value from muonic spectroscopy is shown (green, dashed). The PRad data and fit [4] are shown in blue. The lilac dashed line labeled "Arrington 07" is a fit to pre-Mainz data [5]. The teal dot-dashed line is from [6], a dispersively improved effective-field-theory calculation with the radius as the only free parameter, chosen here to be the value from muonic spectroscopy. The general agreement between the Mainz and earlier data, the PRad data, and this calculation is poor, raising questions about the reliability of all existing form factor measurements.

form factor shape would therefore be another connection to atomic physics. The sensitivity of current data to G_M as well as the possible reach of an experiment at ARIEL are shown in Fig. 2.

The large beam current possible at ARIEL would permit a dramatically thinner target while maintaining sufficient luminosity, even at backward angles. As part of the MAGIX program, the A1 and MAGIX collaborations are developing a jet target [8] which would realize such a thin target, potentially without any background from cell walls or similar. Additionally, the beam-jet interaction region is essentially point-like, reducing systematic uncertainties stemming from acceptance shifts for extended targets and from external radiation.

2.1.2. Stable Nuclei In addition to a proton target, it is possible to extend such a program to light nuclei such as ^{2}H, ^{3}H, ^{3}He, ^{4}He and lithium—nuclei which are also the focus of muon spectroscopy, with some results already available (see [9] and references therein) and others planned. This would open the way to additional cross checks between spectroscopic and accelerator-based measurements.

Additionally, measurements of low-Q^2 form factors for solid materials like carbon are almost trivial, but could improve the world data set considerably.

2.2. Electron Scattering from Radioactive Nuclei

The production of radioactive isotopes and the study of their structures and reactions are the primary objectives of TRIUMF's facility. With its 300 kW superconducting electron driver, the ARIEL facility will expand the isotope production capabilities and further increase its research

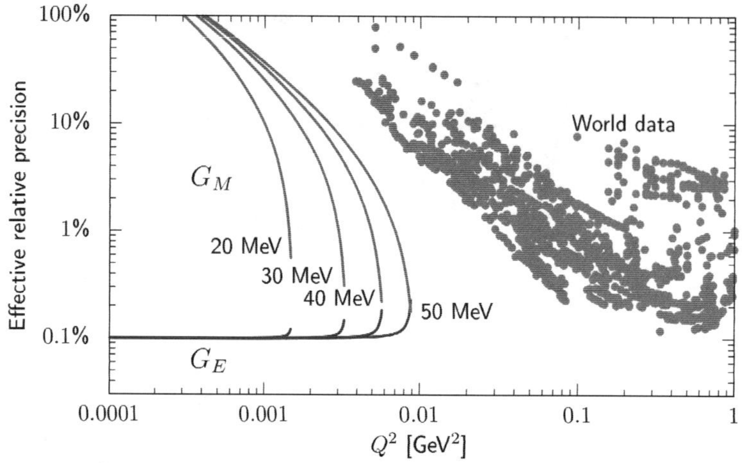

Figure 2. Effective relative precision of the form factors from cross section measurements. Teal (G_M) and red (G_E) lines show the reach and sensitivity for DarkLight@ARIEL assuming cross section measurements on the 0.2% level. Blue points represent the precision on G_M for existing data [7]. Precise data on the form factors at Q^2 below 0.1 GeV^2 are required to extract a reliable radius with minimal model dependence.

opportunities at TRIUMF.

In addition to isotope production, there is the potential to develop a program to perform electron scattering on these isotopes, which would take full advantage of ARIEL's powerful electron driver with TRIUMF's ISOL (Isotope Separation On-Line) facilities and to differentiate research opportunities in the field of nuclear physics from the other radioactive-isotope (RI) facilities, especially fragmentation facilities, in operation and under construction worldwide.

Although such electron scattering work has played an essential role in structure studies of atomic nuclei, it has never been applied to exotic nuclei to date. Studies of the structure and reaction of exotic isotopes have, without exception, been conducted with strongly-interacting particles.

Discussions on the possible application of electron scattering to radioactive nuclei started in the late 1990s [10], when powerful heavy-ion accelerator facilities began to be planned, assuming a collider scheme consisting of an electron ring and a large-scale storage ring to store an energetic radioactive-nucleus beam.

The world's first electron scattering facility dedicated to short-lived RIs, the SCRIT facility at RIKEN in Japan, is in operation today. It is not a collider, but instead consists of an electron storage ring and an ISOL, based on a novel ion-trapping technique named SCRIT (Self-Confining Radioactive-isotope Ion Target) [11]. The SCRIT technique achieves a high luminosity, $\sim 10^{27}\text{cm}^{-2}\text{s}^{-1}$ with a very small number of ions, typically 10^9 [12], whose luminosity is comparable with those expected at the collider projects. This luminosity allows measuring the charge form factor of medium-heavy nuclei covering the first diffraction maximum, and one can determine, in addition to their charge radii, the size and shape of exotic nuclei [13]. For more information on the facility, see T. Suda's contribution to these proceedings.

The radioactive isotopes at the SCRIT facility are produced at an ISOL for the photo-fission reaction of ^{238}U, but the driver beam power will be limited to a maximum of a few kW due to regulations. This SCRIT scheme looks quite compatible with the ARIEL facility, and the ARIEL driver power is more than two orders of magnitude higher.

If an electron storage ring with a few 100 MeV electron beam energy is equipped at ARIEL, the powerful SC electron driver of ARIEL will significantly expand accessible RIs for electron scattering compared with those at the SCRIT facility.

2.3. Photonuclear Reactions for Radioactive Isotopes

Photonuclear reactions provide a way to study the structure of nuclear excited states with well-understood transition operators. Indeed, the photo-absorption cross sections so far available for many stable nuclei [14] provided crucial information to the current understanding of the structures of excited states including collective excitation dynamics.

For radioactive nuclei, the Coulomb excitation in a nucleus-nucleus collision at a forward angle has been the only way so far to study their photonuclear responses. Unfortunately, its interpretation has been always questioned due to the fact that it is not a purely electromagnetic reaction.

Installing an electron scattering facility as discussed above at ARIEL would also be the first to make possible the measurement of the total photoabsorption cross section of radioactive nuclei with a purely electromagnetic probe. As is well known, the inelastic electron-scattering cross section at the ultra-forward angle, $\theta \sim 0°$, is related to the photo-nuclear cross section by virtual-photon theory [15], exactly in the same manner as the Coulomb excitation in a nucleus-nucleus collision.

Detecting scattered electrons that have lost energy in inelastic electron scattering off exotic nuclei at a very forward angle allows for determining the total photo-absorption cross section. One can cover fully the Giant Dipole Resonance (GDR) region with an electron beam energy of $\sim$ 100 MeV [13] under the luminosity of $\sim 10^{27}\text{cm}^{-2}\text{s}^{-1}$, which is required for elastic electron scattering as mentioned above. It should be stressed that this is not possible in a nucleus-nucleus collision even using $\sim$ GeV/u energetic RI beams.

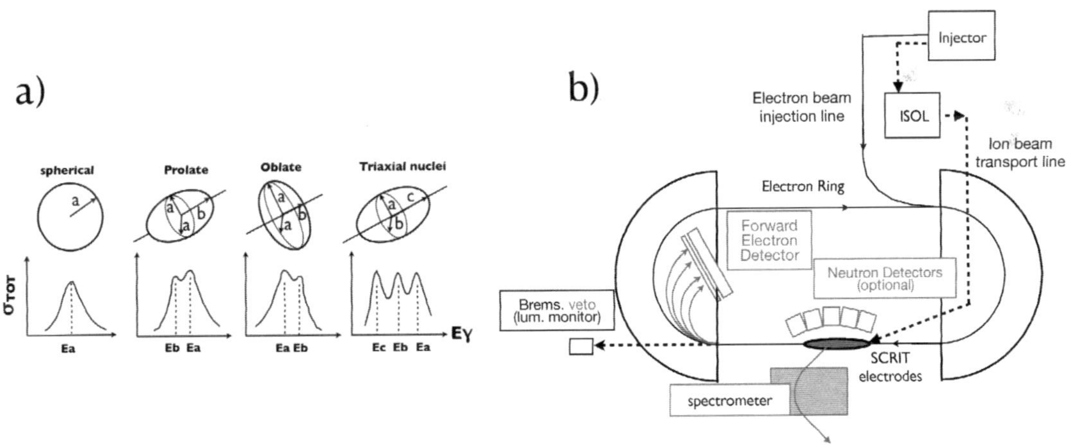

Figure 3. a) Photoabsorption cross section in the GDR region for nuclei having ground-state deformation. b) A conceptual drawing of the future upgrade of SCRIT facility for photonuclear reaction measurements of radioactive nuclei [13]. Those with the black lines indicate existing facilities and the red lines required equipment.

One physics case-study at such a facility is shown in Fig. 3 a). Photonuclear responses of exotic nuclei in the GDR region may reveal exotic deformations in the ground state including the triaxial deformation [16]. This idea has been discussed as one of the possible future upgrades of the SCRIT facility as shown in Fig. 3 b). Such studies could potentially be adapted to serve broader purposes, such as searches for permanent electric dipole moments in atomic nuclei, where detection of a nonzero result would imply sources of CP violation beyond the Standard Model. The sensitivity of these searches would be enhanced by the appearance of pear-shaped deformation in the ground states of exotic nuclei [17, 18, 19, 20, 21]. Radioactive beam studies at ISOLDE, e.g., have used Coulomb excitation [17, 19] and other methods [18, 20] to help identify the best-suited nuclei for such studies, and electron-scattering studies could potentially play an important complementary role.

2.4. Nuclear Astrophysics
Radiative capture reactions, i.e. nuclear reactions in which the incident projectile is absorbed by the target nucleus and γ radiation is then emitted, play a crucial role in nucleosynthesis processes in stars [22], hence knowledge of their reaction rates at stellar energies is essential to understanding the abundance of the chemical elements in the universe. However, determination of these reaction rates has proven to be challenging, principally due to the Coulomb repulsion between initial-state nuclei and the weakness of the electromagnetic force. For example, the decay of unbound nuclear states by the emission of a particle of the same type as that captured, or by the emission of some other type of particle, is often 10^3–10^6 times more probable than decay by γ emission.

The availability of an intense, low-energy electron beam rather than a photon beam allows the possibility to measure the inverse reaction. This idea has been previously proposed but not measured, and was more recently considered in detail [23] for the case of $C^{12}(\alpha,\gamma)^{16}O$. See contribution by I. Friščić in this proceedings. In addition to a low-energy electron beam, a gas jet target, high resolution magnetic spectrometer and α recoil detector are required to carry out the experiment.

3. QED Tests
3.1. Threshold Positron Production
In both pair and triplet production a positron and electron are produced spontaneously as a photon interacts with a strong electric field from either a nucleus (pair production) or an electron (triplet production) as shown in Fig. 4.

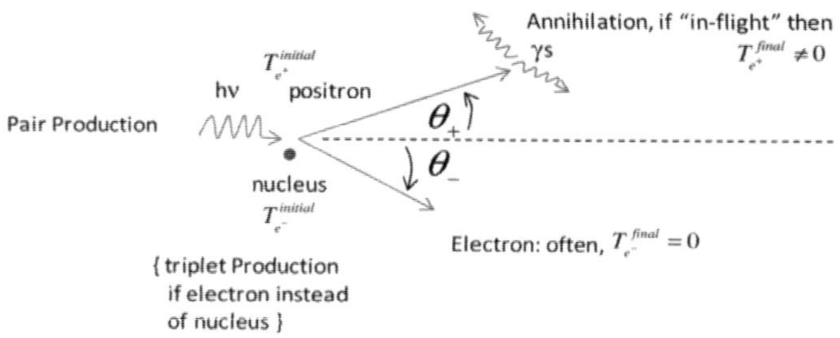

Figure 4. Pair and triplet e^+e^- production from an atom.

The threshold electron beam energy for these effects to take place is

$$(h\nu)_{\min} = 2m_0c^2\left(1+\frac{m_0}{M}\right) , \tag{3}$$

where m_0 and M are the incident electron and target masses, respectively and

$$M \equiv M_{\text{nucleus}} >> m_0 \text{ (pair production)} \tag{4}$$
$$M = m_0 \text{ (triplet production)} . \tag{5}$$

The mean kinetic energy given to each of the two particles is half of the available kinetic energy T_{avail} (actually the positron gets a bit more energy because of the push from the positively charged nucleus). The mean angle given to each of the two particles with respect to the incident electron directions is

$$\bar{\theta}_{\pm} = \frac{2m_0c^2}{T_{avail}} \ . \tag{6}$$

The $1/T$ dependence is similar to bremsstrahlung: higher energy particles get more forward directed.

Bethe and Heitler [24] derived the cross section per atom for **pair production** as

$$\frac{d\kappa_a^{pair}}{dT_{\pm}} = \sigma_0 \frac{Z^2}{T_{avail}} P \ , \tag{7}$$

where P is a function shown in Fig. 5, Z is the nuclear charge and $\sigma_0 \equiv \frac{r_0^2}{137} = 4.8 \times 10^{-28}\text{cm}^2/\text{atom}$.

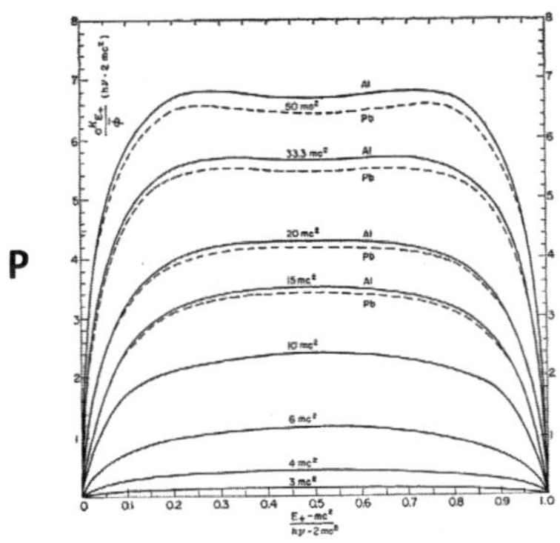

Figure 5. P vs. kinetic energy fraction given to the positron from [25]. Notice the symmetry: energy not given to the positron is given to the electron and vice-versa.

In the case of **triplet production**, the electric field is now from an electron, a very light particle, which becomes indistinguishable from the created particle. The mean kinetic energy given to each of the three particles is now one-third of the available kinetic energy. The threshold is now $4m_0c^2$. The triplet production cross section is related to the pair production cross section as

$$\kappa_a^{triplet} = \kappa_a^{pair} \cdot \frac{1}{CZ} \ , \tag{8}$$

where $C \approx 1$ and has no Z dependence.

Measurements have been reported [27, 28, 29, 30] with sporadic sightings of triplet events. The first definitive observation of triplet production was in 1972 [26] using a streamer chamber (Fig. 6). At ARIEL, such an experiment could be mounted as well. With 2 μA of 2.5 MeV electron beam on a 5 micron thick carbon foil (luminosity = 3×10^{33} cm^{-2} s^{-1}) the rate of positrons produced via the triplet mechanism is estimated (using [26]) at about 2 Hz into a spectrometer of solid angle 1 msr. This should allow much more precise measurements of positrons at threshold than previously possible. Clearly, low electron beam energies of order 1-3 MeV are required.

3.2. Radiative Møller Scattering

Møller (electron-electron) scattering is a background in electron scattering experiments, and is a purely QED process at low energies. It is theoretically straightforward to calculate, and has been so for decades. However, a modern retrospective has revealed gaps in previous treatments,

Journal of Physics: Conference Series **2391** (2022) 012001 doi:10.1088/1742-6596/2391/1/012001

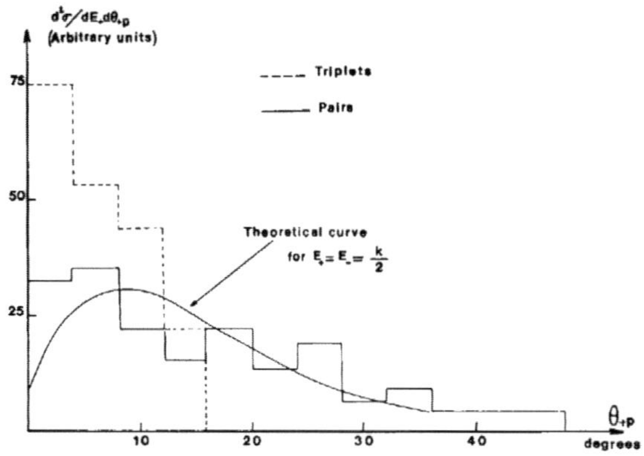

Figure 6. Positron angular distribution relative to the projected emission angle from [26]. Experimental histograms are given for triplet and pair production. The theoretical curve is plotted for pair production.

particularly the omission of the electron mass in the calculation of the radiative diagrams shown in Fig. 7. Corrections due to these diagrams are typically included as a multiplicative factor to the Born cross section:

$$\frac{d\sigma}{d\Omega}|_{\rm soft} = (1+\delta)\frac{d\sigma}{d\Omega}|_{\rm Born}\ , \tag{9}$$

with $\delta = \delta(\Delta E, \Omega)$. This traditional method requires defining a cut-off ΔE, the maximum amount of energy a photon can carry away for which the event passes acceptance cuts. For an experiment having spectrometers with small, well-defined energy and angular acceptances, this formulation of the radiative corrections can be applied straightforwardly.

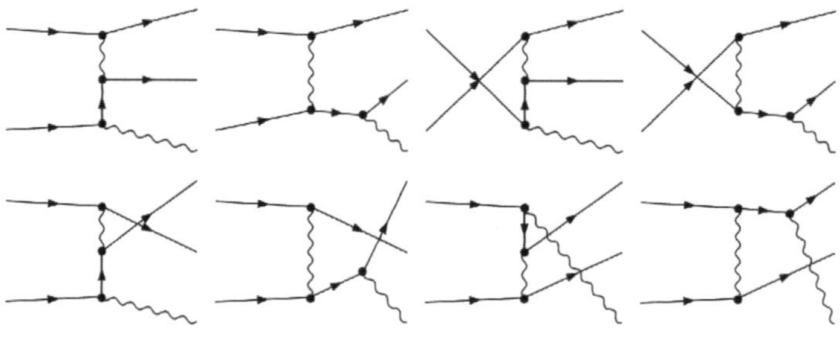

Figure 7. Feynman diagrams for radiative Møller scattering.

Radiative corrections for hard-photon bremsstrahlung emission in both Møller and Bhabha scattering have been performed [31] in a consistent approach without ultra-relativistic approximations and permit a complete analysis at next-to-leading-order.

Measurement of radiative Møller scattering as a function of energy between 10 and 100 MeV is highly desirable to validate the correction procedure. This has been carried out at 2.5 MeV (where radiative effects are negligible) using a focusing spectrometer [32] by the DarkLight collaboration. The desired measurements can be straightforwardly carried out using the ARIEL electron linac with the apparatus proposed for the DarkLight experiment.

4. Searches for Dark Matter and BSM Physics

A dark sector search is already being undertaken at the ARIEL e-linac by the DarkLight collaboration, so it is natural to explore related searches that might be done with the same

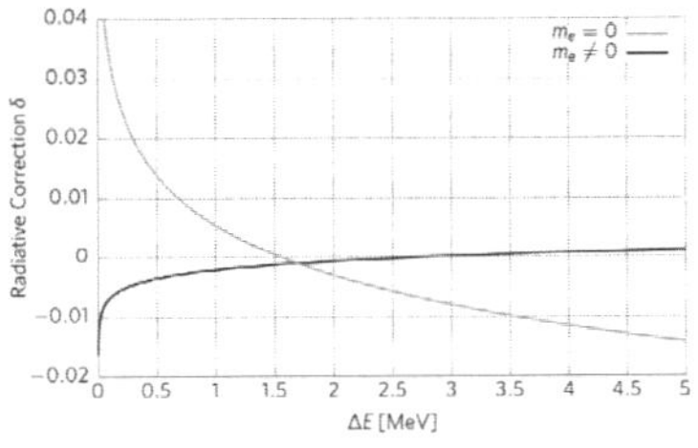

Figure 8. Comparison of the Møller radiative correction term δ for a 100 MeV electron beam at 5° in the CM frame for different electron masses from [31].

accelerator. The physics scope for the ARIEL e-linac is limited by its maximum energy of 50 MeV, and it is important to note that the low-energy dark sector landscape in both visible and invisible final states is well populated by a range of existing experiments. Finding a new area where an experiment at TRIUMF can make a strong contribution depends on either identifying an uncovered niche outside of the benchmark simplified models typically used by the field, as DarkLight did with protophobic bosons [33], or by taking advantage of the ARIEL e-linac's unique strength: the high intensity of its beam. The clearest opportunities presented are, e.g. rare processes at low mass which are too time-consuming to probe with lower intensity beams, or for which the low energy allows for some particular access to challenging kinematics or limits SM backgrounds. In what follows we offer concrete examples that illustrate the possibilities.

4.1. Delbrück Scattering to Search for Dark Photons of sub-MeV in Mass

A dark photon, the boson A' associated with a U(1) gauge symmetry of a dark sector [34], can mix with the SM photon with strength ε, with $\varepsilon \ll 1$, to produce millicharged couplings, namely, $\mathcal{L}_{\rm mix} = \varepsilon A'_\mu J^\mu_{\rm EM}$, with the electromagnetically charged fermions of the SM. The nature of this "portal" interaction between the dark and visible sectors allows an enormous range of dark photon masses and mixing parameters to appear, and we must look to experiment to constrain the possibilities. Studies in e^-e^+ collisions [35, 36] have been very effective in limiting the parameter space for dark photons in excess of $2m_e$ in mass, and there are a broad sweep of astrophysical and cosmological constraints as well [37, 38]. However, if the dark photon mass $m_{A'}$ satisfies $m_{A'} \lesssim 2m_e$, then its mixing with the photon guarantees that the decay $A' \to 3\gamma$ can occur, though the dark photon in this case can be quite long-lived, because the decay is a one-loop process with a rate that scales as $\varepsilon^2\alpha^4$.

Nevertheless, there may be discoverable phase space yet to explore [39]. This conclusion largely stems from a proper assessment of the decay rate: Earlier work computed the decay rate in the Euler-Heisenberg limit [40], which assumes that $m_{A'} \ll m_e$, but the exact one-loop calculation yields a result some 10-100 times larger if $850\,\mathrm{keV} \lesssim m_{A'} \lesssim 1\,\mathrm{MeV}$, shortening the dark photon lifetime and altering the excluded phase space significantly [39]. Referring to Fig. 4 of [39], we note that $850\,\mathrm{keV} \lesssim m_{A'} \lesssim 1\,\mathrm{MeV}$ and $10^{-5} \lesssim \varepsilon' \lesssim 10^{-4}$ represents an interesting window of opportunity [39], with constraints coming from big-bang nucleosynthesis, which would appear to demand $\tau_{A'} < 1\,\mathrm{sec}$, and from the anomalous magnetic moment of the electron, $(g-2)_e$ [41, 42]. A constraint also comes from nonobservation of $A' + e \to e + \gamma$ in the LSND experiment, presuming the production of A' from $\pi^0 \to A'\gamma$ decay [43]. The authors of [43] thus contend this region is excluded, but the LSND constraint can be evaded through a non-minimal dark vector model, such as the protophobic gauge boson. We note, too, the experimental controversy in the precise value of the fine-structure constant α from atom interferometry [44, 45], noting [46] for context, with resolutions that would alter the $(g-2)_e$ constraint of [42] somewhat.

Delbrück scattering, in which a photon is deflected in a strong Coulomb field due to vacuum polarization, would be sensitive to the A' in this region without assumption concerning the

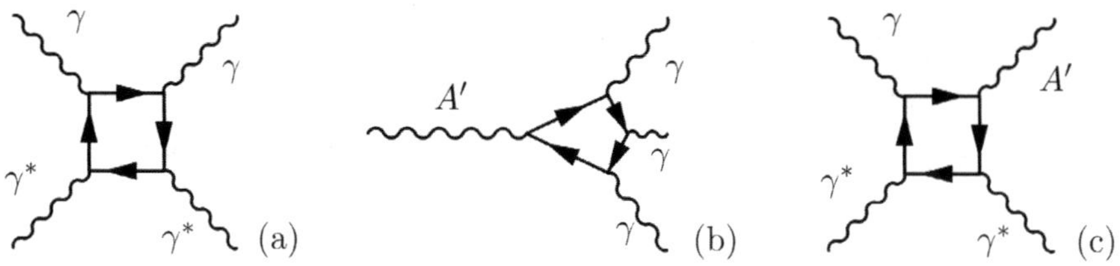

Figure 9. Representative Feynman diagrams to illustrate (a) Delbrück scattering, which is light scattering in the Coulomb field of an atom or nucleus, as indicated by the appearance of γ^*, (b) $A' \to 3\gamma$ decay, and (c) Delbrück scattering modified to probe a sub-MeV dark photon, with the appearance of missing momentum giving the experimental signature. Note that crossed diagrams exist but have not been shown. The solid lines denote any of the electrically charged fermions of the SM, with the electron/positron giving the dominant contribution.

production mechanism. Pertinent diagrams are illustrated in Fig. 9. We note that at ARIEL energies, Delbrück scattering has been clearly identified [47]. The suggested new-physics search would require the ability to assess the existence of missing momentum or energy.

4.2. $|\Delta B| = 2$ *Processes in Nuclei*

Experimental limits on processes that would change baryon number by one unit, i.e., $|\Delta B| = 1$, are among the most stringent known to science [48, 49]. Experimental studies on processes that would change baryon number by two or more units, however, have been much more limited [50], although their physical origin can be altogether different [51, 52, 53]. There is clear reason to search for such effects: The existence of $|\Delta(B-L)| = 2$ violation is necessary to make a massive neutrino its own antiparticle [54], and observable B and/or $B-L$ violation in the quark sector would help to identify its dynamical mechanism.

The ARIEL accelerator facility's intense electron beam would be well suited to the search for $|\Delta B| = 2$ through low-energy channels. Of particular note are the processes $e^- p \to e^+ \bar{p}$ or $e^- p \to \bar{n}\bar{\nu}$ [53], with the $B-L$-violating channel $e^- n \to e^- \bar{n}$ also open to study [55] — here the use of a deuteron target could prove advantageous [55, 53]. The estimated event rates for the B violating, but $B-L$ conserving, processes we have noted are no more than $\mathcal{O}(10)$ events per year [56, 57]. Thus the low energy of the existing electron linac at ARIEL is a key advantage, rather than a limitation, both in controlling backgrounds and in detecting the produced anti-nucleons. At ARIEL, backgrounds are controlled by the extremely low energy of the electron scattering process; in particular, the accelerator operates at energies far below pion production threshold. Moreover, the produced anti-nucleon would be at sufficiently low energy that the annihilation should be prompt and thus occur within the target, typically yielding a five-pion final state, as determined in R&D studies of searches for $n - \bar{n}$ oscillations [58, 59]. In the ARIEL environment, the supposed five-pion final state signal should be quite striking. Their total charge, in a proton target, should signal the electric charge of the anti-nucleon and hence the precise process. This search could potentially run parasitically at ARIEL.

4.3. Beam-Dump Experiments to Search for Dark Matter

The high intensity of the ARIEL e-linac makes it an interesting potential location to consider a beam dump dark matter experiment as well. The sensitivity relative to DarkMESA, LDMX, and other light DM searches should be examined using an appropriate benchmark model and the usefulness of such an experiment determined in light of the fact that it would necessarily begin much later than already-approved experiments.

Dark matter searches at beam dumps are relatively model-independent. A high-current beam is directed to a target where dark sector particles may be created. Enough shielding to stop essentially all Standard Model particles is placed behind or incorporated into the target, and a detector behind the shielding searches for dark matter particles exiting the dump. The detector is sensitive to dark matter particles via their scattering in the target material. Examples of such experiments, interpretations, and proposals include BDX [60], DarkMESA (see contribution by L. Doria in this proceedings), E137 [61], the SHiP Scattering Neutrino Detector [62], COHERENT [63], and MiniBooNE (in its dedicated dark matter configuration [64]). A schematic of a beam dump DM experimental setup, including a simplified detector of moderate scale, is shown in Fig. 10.

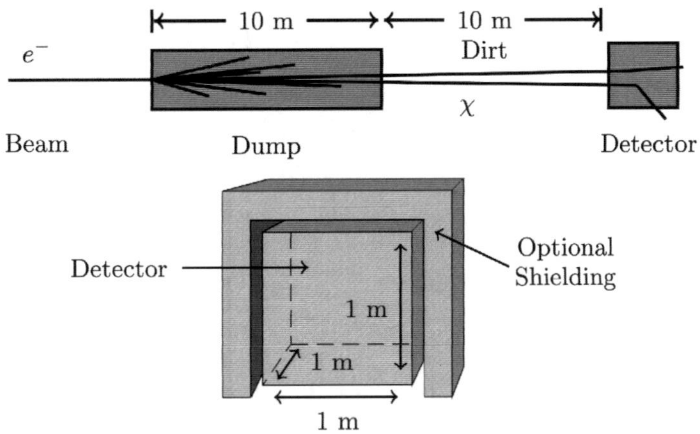

Figure 10. Schematic of a beam dump experiment searching for light dark matter, from [65].

Beam-dump experiments probe the same types of final states as missing-momentum or missing-energy based experiments (LDMX, NA64). They share with them the advantage that these approaches, unlike visible final state searches, are relatively agnostic to the lifetime of the mediator particle(s) and to the decay chain producing the dark matter and can therefore constrain a wide range of models [66]. Scattering-based beam dump experiments are, however, disadvantaged relative to missing-momentum experiments in that they depend on an additional interaction and therefore gain a $\sim \alpha_D \epsilon^2$ suppression factor [67]. The difference could be made up with sufficient beam intensity, since missing momentum experiments must run at low current, and the ARIEL e-linac's uniquely high power makes it worth investigating as a potential site.

A viable location for a beam dump experiment does exist at the ARIEL e-linac. A small room behind the ARIEL targets could host a detector, allowing this new experiment to take data parasitically to ARIEL and without disrupting the operation of DarkLight or another future experiment in its position. One potential experimental challenge is the low momentum transfer from scattering DM particles in the detector at the mass and momentum ranges accessible at ARIEL. This is not insurmountable but would require thoughtful detector design.

The current and proposed future exclusion landscape for light dark matter in the context of invisible dark photon decays is shown in Fig. 11. The dominant sensitivity in the sub-10 MeV mass range is projected to come from LDMX. For reference, the landscape for visible dark photon decays is shown in Fig. 12.

It could also be considered whether there is any scope for a missing-energy or missing-momentum search at the ARIEL e-linac. The practical advantages of a classic beam dump experiment at TRIUMF would be negated, since such an experiment could not operate

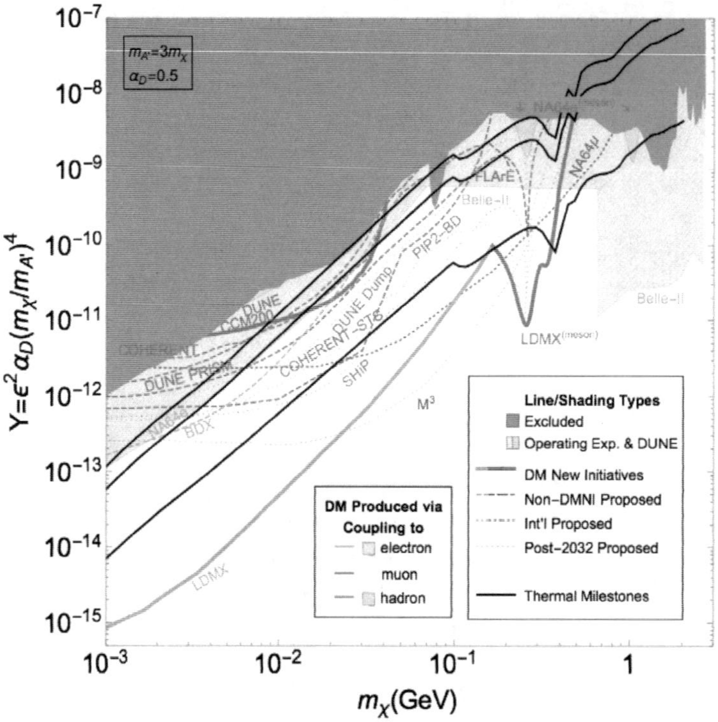

Figure 11. Constraints on light dark matter in the context of invisible decays of dark photons from current, future, and proposed experiments [68].

parasitically behind the ARIEL targets, but it would offer the opportunity for significantly higher sensitivity. This approach would require a solution to the experimental challenges of missing momentum reconstruction in the presence of extremely high electron multiplicity in the beam bunches.

We conclude this section by noting another way in which dark-sector searches at the ARIEL facility could be complementary to accelerator-based DM experiments planned elsewhere. At the proposed LDMX experiment [66, 67, 70], e.g., invisible decays of a dark photon A' can be probed through its missing-momentum signature. That is, as shown in Fig. 13, an A' produced through bremsstrahlung in electron-nucleus scattering can decay invisibly via $A' \to \chi\chi$, and this possibility can be probed through the measurement of the momentum of the final-state electron. This may be a unique possibility to probe χ because χ carries no SM charges. However, in place of the usual dark photon, the gauge boson mediator could couple to baryon number instead [71, 72, 73, 49]. The gauge boson in such a model couples to neutrons and photons, and it can couple to electrically charged particles, through kinetic mixing with the SM photon. In this case the dark gauge boson can still be produced via bremsstrahlung, but it could decay to a $\chi\bar{\chi}$ pair, in which χ carries baryon number $B = 1/3$. Figure 13 also illustrates this possibility upon the replacement of $\chi\chi$ with $\chi\bar{\chi}$.

Although this model would generate a missing momentum signature, χ and $\bar{\chi}$ could also interact with neutrons and protons in the far detector. One possibility with a striking observable signature concerns *destabilization* of the nucleus. Here the incoming DM particle χ stimulates the decay of a nucleus in the far detector via the neutron-decay process $\bar{\chi}n \to \chi\chi$. We note that $n \to 3\chi$ has also been considered in the context of the neutron lifetime anomaly [74], noting [75], and the decay rate need not be exceedingly small [49]. The breakup of the nucleus, with

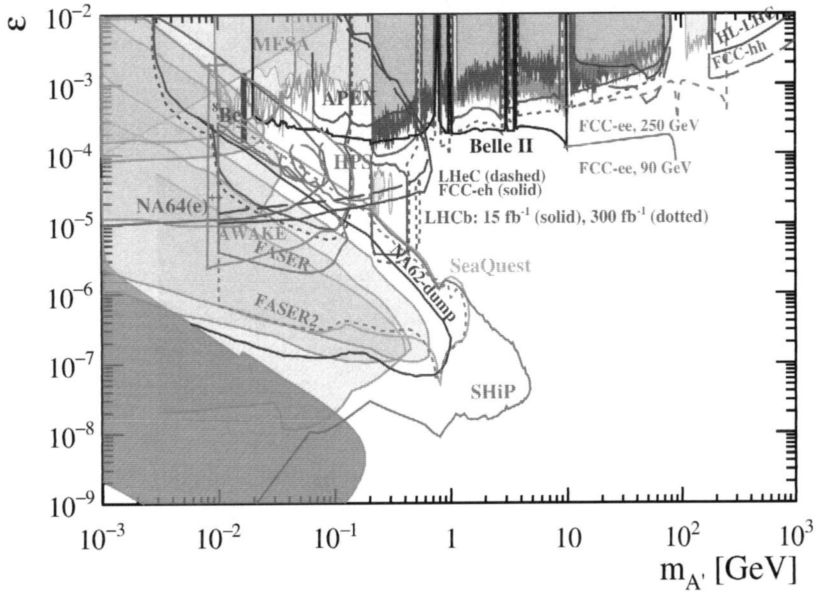

Figure 12. Constraints on dark photons detectable through visible decays in the context of current, future, and proposed experiments. Grayed regions represent existing limits, in the assumption of flavor-independent couplings. Colored lines represent future projected exclusions [69].

roughly a GeV of energy loss, could be documented through gamma detectors placed at the far detector. The latter could be sufficiently removed from the electron beam environment to make such a photon fingerprinting process possible. The viability of such an approach has been studied in work by KamLAND [76], and references therein.

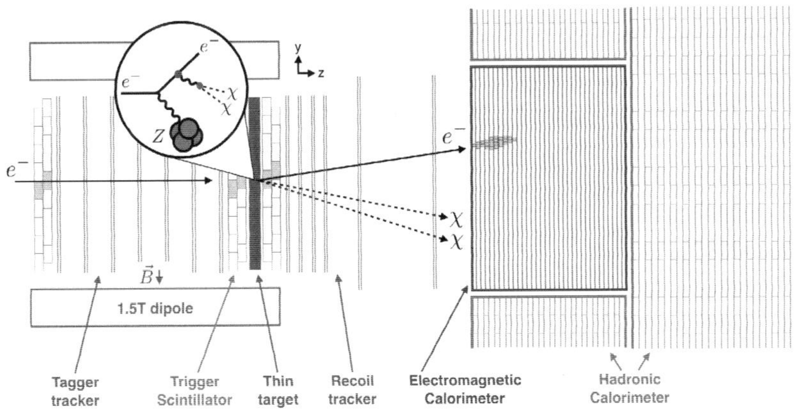

Figure 13. Illustration of the production of dark matter particles χ through dark photon decay, apropos to the LDMX detector, from [70]. Replacing $\chi\chi$ with $\chi\bar{\chi}$ illustrates the possibility of a dark gauge boson coupled to baryon number, as discussed in the text.

4.4. Search for X17 in e^+e^- Decay of Giant Dipole Resonance
It has been suggested that the X17 may also be able to be produced in the decay of giant nuclear resonances. Both Isoscalar Giant Monopole Resonances (ISGMR) and Isovector Giant Dipole Resonances (IVGDR) can be readily excited in an electron beam fixed target experiment, and the typical energy scale of these resonances (ISGMR) follows Eq. 10. For atomic numbers above about 20, these can be readily reached with a 30-50 MeV beam.

$$E_c = 80A^{-1/3}\text{MeV} \tag{10}$$

Final states may carry angular distributions associated with the excited multipole but should be randomly oriented in the detector frame, producing a signature that is independent of the angle with respect to the electron beam, distinguishing those decay particles from the forward-peaking prompt events.

Unfortunately, electron scattering has a large contribution from the radiative tail, which increases the background associated with identifying a resonance, and identifying the particles from that resonance's decay, substantially. At the collision energies of the current or upgraded ARIEL accelerator, techniques to separate prompt and resonant final states spatially or temporally would need to have exquisite resolutions that are not currently feasible. The lifetime of the resonance is extremely short (typically 10^{-21} to 10^{-19} s), and its recoil energy essentially zero. At higher beam energies these constraints are relaxed, but these are beyond the upgrades currently discussed for ARIEL.

We note that different excitation techniques, in particular tagged photons, may make giant resonances a more fruitful area both for exotic particle searches and other studies of resonance structures, but would require the development of a photon-tagging station on the beamline.

5. Positron Source for Materials Science Research
Low energy positrons are a unique probe of materials and surfaces that allow measurements of surface properties not possible with other probes such as electrons or X-rays. Though nuclear decay sources are available for positron studies, accelerator sources can provide higher fluxes and much higher brightness. See [77] for a detailed consideration of such a source at the Low Energy Recirculator Facility at Jefferson Laboratory, Newport News, VA, USA.

ARIEL upgraded in energy and operating with energy recovery has the potential to produce a positron source of great interest for materials science research. The addition of spin-polarization capability to the positron source would enhance the research scope. Full realization of such a capability would likely attract a high demand from users.

6. Summary

A high-intensity electron beam at low energies as can be provided at ARIEL offers the potential to pursue a suite of important scientific questions. We have presented many here, along with initial exploration of the challenges and potential reach for each. Several require significant investment to realize more sophisticated capabilities. Further study will be essential to determine whether the investment is merited.

References

[1] Randolf Pohl, Aldo Antognini, Francois Nez, Fernando D. Amaro, Francois Biraben, Joao M. R. Cardoso, Daniel S. Covita, Andreas Dax, Satish Dhawan, Luis M. P. Fernandes, Adolf Giesen, Thomas Graf, Theodor W. Hansch, Paul Indelicato, Lucile Julien, Cheng-Yang Kao, Paul Knowles, Eric-Olivier Le Bigot, Yi-Wei Liu, Jose A. M. Lopes, Livia Ludhova, Cristina M. B. Monteiro, Francoise Mulhauser, Tobias Nebel, Paul Rabinowitz, Joaquim M. F. dos Santos, Lukas A. Schaller, Karsten Schuhmann, Catherine Schwob,

David Taqqu, Joao F. C. A. Velos, and Franz Kottmann. The size of the proton. *Nature*, 466:213–216, 2010.

[2] J. C. Bernauer, P. Achenbach, C. Ayerbe Gayoso, R. Bohm, D. Bosnar, L. Debenjak, M. O. Distler, L. Doria, A. Esser, H. Fonvieille, J. M. Friedrich, J. Friedrich, M. Gomez Rodriguez de la Paz, M. Makek, H. Merkel, D. G. Middleton, U. Muller, L. Nungesser, J. Pochodzalla, M. Potokar, S. Sanchez Majos, B. S. Schlimme, S. Sirca, Th Walcher, and M. Weinriefer. High-precision determination of the electric and magnetic form factors of the proton. *Phys.Rev.Lett.*, 105:242001, 2010.

[3] Peter J. Mohr, Barry N. Taylor, and David B. Newell. CODATA recommended values of the fundamental physical constants: 2006. *Rev. Mod. Phys.*, 80(2):633–730, 06 2008.

[4] W. Xiong et al. A small proton charge radius from an electron–proton scattering experiment. *Nature*, 575(7781):147–150, 2019.

[5] J. Arrington, W. Melnitchouk, and J. A. Tjon. Global analysis of proton elastic form factor data with two-photon exchange corrections. *Physical Review C*, 76(3):035205, 09 2007.

[6] J. M. Alarcón, D. W. Higinbotham, C. Weiss, and Z. Ye. Proton charge radius extraction from electron scattering data using dispersively improved chiral effective field theory. *Phys. Rev. C*, 99(4):044303, 2019.

[7] J. C. Bernauer, M. O. Distler, J. Friedrich, Th. Walcher, P. Achenbach, C. Ayerbe Gayoso, R. Böhm, D. Bosnar, L. Debenjak, L. Doria, A. Esser, H. Fonvieille, M. Gómez Rodríguez de la Paz, J. M. Friedrich, M. Makek, H. Merkel, D. G. Middleton, U. Müller, L. Nungesser, J. Pochodzalla, M. Potokar, S. Sánchez Majos, B. S. Schlimme, S. Širca, and M. Weinriefer. Electric and magnetic form factors of the proton. *Phys. Rev. C*, 90:015206, 07 2014.

[8] B. S. Schlimme et al. Operation and characterization of a windowless gas jet target in high-intensity electron beams. *Nucl. Instrum. Meth. A*, 1013:165668, 2021.

[9] Aldo Antognini, Franz Kottmann, and Randolf Pohl. Laser spectroscopy of light muonic atoms and the nuclear charge radii. *SciPost Phys. Proc.*, 5:021, 2021.

[10] T. Katayama, T. Suda, and I. Tanihata. Status of MUSES project and electron RI collider at RIKEN. In *Physica Scripta T*, volume 104, 2003.

[11] M. Wakasugi, T. Suda, and Y. Yano. A new method for electron-scattering experiments using a self-confining radioactive ion target in an electron storage ring. *Nuclear Instruments and Methods*, 532(1-2), 2004.

[12] K. Tsukada, A. Enokizono, T. Ohnishi, K. Adachi, T. Fujita, M. Hara, M. Hori, T. Hori, S. Ichikawa, K. Kurita, K. Matsuda, T. Suda, T. Tamae, M. Togasaki, M. Wakasugi, M. Watanabe, and K. Yamada. First Elastic Electron Scattering from Xe 132 at the SCRIT Facility. *Physical Review Letters*, 118(26):1–5, 2017.

[13] Toshimi Suda and Haik Simon. Prospects for electron scattering on unstable, exotic nuclei. *Progress of Particle and Nuclear Physics*, 96:1–31, sep 2017.

[14] B L Berman and S C Fultz. Measurements of the giant dipole resonance with monoenergetic photons. *Reviews of Modern Physics*, 47(3):713–761, 1975.

[15] C A Bertulani. Excitation of soft dipole modes in electron scattering. *Phys. Rev. C*, 75(2):24606, feb 2007.

[16] R. Berger. Features of the giant e1 resonance. In Carlo Schaerf Sergio Costa, editor, *Photo Nuclear Reaction, Lecture Notes in Physics*, chapter 3, pages 1–222. Springer-Verlag, 1977.

[17] L. P. et al. Gaffney. Studies of pear-shaped nuclei using accelerated radioactive beams. *Nature*, 497(7448):199–204, 2013.

[18] P. A. Butler et al. The observation of vibrating pear-shapes in radon nuclei. *Nature Commun.*, 10(1):2473, 2019.

[19] P. A. Butler et al. Evolution of Octupole Deformation in Radium Nuclei from Coulomb Excitation of Radioactive ^{222}Ra and ^{228}Ra Beams. *Phys. Rev. Lett.*, 124(4):042503, 2020.

[20] M. M. R. et al. Chishti. Direct measurement of the intrinsic electric dipole moment in pear-shaped thorium-228. *Nature Physics*, 16(8):853–856, 2020.

[21] J. A. Behr. Nuclei with enhanced Schiff moments in practical elements for atomic and molecular EDM measurements. 3 2022.

[22] Carl R. Brune and Barry Davids. Radiative capture reactions in astrophysics. *Annual Review of Nuclear and Particle Science*, 65(1):87–112, 2015.

[23] I. Friščić, T. W. Donnelly, and R. G. Milner. New approach to determining radiative capture reaction rates at astrophysical energies. *Phys. Rev. C*, 100:025804, Aug 2019.

[24] H. Bethe and W. Heitler. On the Stopping of Fast Particles and on the Creation of Positive Electrons. *Proceedings of the Royal Society of London Series A*, 146(856):83–112, August 1934.

[25] Charlotte Meaker Davisson and Robley D. Evans. Gamma-ray absorption coefficients. *Rev. Mod. Phys.*, 24:79–107, Apr 1952.

[26] G. Roche, J. Arnold, J. Augerat, L. Avan, M. Avan, J. Bonnet, J. Fargeix, A. Fleury, J. Jousset, M.J. Parizet, and M. Vialle. First results of triplet production measurements near the threshold with a streamer chamber

technique: Experimental set-up. *Nuclear Instruments and Methods*, 103(3):533–544, 1972.

[27] Marques da Silva, Aurelio. Contribution à l'étude de la matérialisation de l'énergie. *Ann. Phys.*, 11(11):504–547, 1939.

[28] Kenichi Shinohara and Mitio Hatoyama. Pair production in the field of an electron. *Phys. Rev.*, 59:461–461, Mar 1941.

[29] W. E. Ogle and P. Gerald Kruger. Pair electrons formed in the field of an electron. *Phys. Rev.*, 67:282–285, May 1945.

[30] J. A. Phillips and P. Gerald Kruger. On the production of electron pairs in the field of an electron. *Phys. Rev.*, 76:1471–1478, Nov 1949.

[31] Charles S. Epstein and Richard G. Milner. Qed radiative corrections to low-energy møller and bhabha scattering. *Phys. Rev. D*, 94:033004, Aug 2016.

[32] C. S. Epstein et al. Measurement of Møller scattering at 2.5 MeV. *Phys. Rev. D*, 102(1):012006, 2020.

[33] Jonathan L. Feng, Bartosz Fornal, Iftah Galon, Susan Gardner, Jordan Smolinsky, Tim M. P. Tait, and Philip Tanedo. Protophobic Fifth-Force Interpretation of the Observed Anomaly in ^{8}Be Nuclear Transitions. *Phys. Rev. Lett.*, 117(7):071803, 2016.

[34] Bob Holdom. Two U(1)'s and Epsilon Charge Shifts. *Phys. Lett. B*, 166:196–198, 1986.

[35] J. P. Lees et al. Search for a Dark Photon in e^+e^- Collisions at BaBar. *Phys. Rev. Lett.*, 113(20):201801, 2014.

[36] J. P. Lees et al. Search for Invisible Decays of a Dark Photon Produced in e^+e^- Collisions at BaBar. *Phys. Rev. Lett.*, 119(13):131804, 2017.

[37] Andrea Caputo, Alexander J. Millar, Ciaran A. J. O'Hare, and Edoardo Vitagliano. Dark photon limits: A handbook. *Phys. Rev. D*, 104(9):095029, 2021.

[38] D. Antypas et al. New Horizons: Scalar and Vector Ultralight Dark Matter. 3 2022.

[39] Samuel D. McDermott, Hiren H. Patel, and Harikrishnan Ramani. Dark Photon Decay Beyond The Euler-Heisenberg Limit. *Phys. Rev. D*, 97(7):073005, 2018.

[40] Maxim Pospelov, Adam Ritz, and Mikhail B. Voloshin. Bosonic super-WIMPs as keV-scale dark matter. *Phys. Rev. D*, 78:115012, 2008.

[41] D. Hanneke, S. Fogwell, and G. Gabrielse. New Measurement of the Electron Magnetic Moment and the Fine Structure Constant. *Phys. Rev. Lett.*, 100:120801, 2008.

[42] Maxim Pospelov. Secluded u(1) below the weak scale. *Phys. Rev. D*, 80:095002, Nov 2009.

[43] Maxim Pospelov and Yu-Dai Tsai. Light scalars and dark photons in Borexino and LSND experiments. *Phys. Lett. B*, 785:288–295, 2018.

[44] Richard H. Parker, Chenghui Yu, Weicheng Zhong, Brian Estey, and Holger Müller. Measurement of the fine-structure constant as a test of the Standard Model. *Science*, 360:191, 2018.

[45] Léo Morel, Zhibin Yao, Pierre Cladé, and Saïda Guellati-Khélifa. Determination of the fine-structure constant with an accuracy of 81 parts per trillion. *Nature*, 588(7836):61–65, 2020.

[46] Susan Gardner and Xinshuai Yan. Light scalars with lepton number to solve the $(g-2)_e$ anomaly. *Phys. Rev. D*, 102(7):075016, 2020.

[47] A. I. Milstein and Martin Schumacher. Present status of Delbruck scattering. *Phys. Rept.*, 243:183–214, 1994.

[48] R. L. Workman and Others. Review of Particle Physics. *PTEP*, 2022:083C01, 2022.

[49] Jeffrey M. Berryman, Susan Gardner, and Mohammadreza Zakeri. Neutron Stars with Baryon Number Violation, Probing Dark Sectors. *Symmetry*, 14(3):518, 2022.

[50] Pavel Fileviez Perez et al. On Baryon and Lepton Number Violation. 7 2022.

[51] Rabindra N. Mohapatra and R. E. Marshak. Local B-L Symmetry of Electroweak Interactions, Majorana Neutrinos and Neutron Oscillations. *Phys. Rev. Lett.*, 44:1316–1319, 1980. [Erratum: Phys.Rev.Lett. 44, 1643 (1980)].

[52] Jonathan M. Arnold, Bartosz Fornal, and Mark B. Wise. Simplified models with baryon number violation but no proton decay. *Phys. Rev. D*, 87:075004, 2013.

[53] Susan Gardner and Xinshuai Yan. Processes that break baryon number by two units and the Majorana nature of the neutrino. *Phys. Lett. B*, 790:421–426, 2019.

[54] Steven Weinberg. Baryon and Lepton Nonconserving Processes. *Phys. Rev. Lett.*, 43:1566–1570, 1979.

[55] Susan Gardner and Xinshuai Yan. Phenomenology of neutron-antineutron conversion. *Phys. Rev. D*, 97(5):056008, 2018.

[56] Susan Gardner and Xinshuai Yan. Prospects for the accelerator-based discovery of baryon-number violation by two units. to appear on the arXiv.

[57] Susan Gardner. New Opportunities for the Study of Baryon-Number Violation at Low-Energy Accelerators. contribution to these proceedings.

[58] D. G. Phillips, II et al. Neutron-Antineutron Oscillations: Theoretical Status and Experimental Prospects.

Phys. Rept., 612:1–45, 2016.

[59] A. Addazi et al. New high-sensitivity searches for neutrons converting into antineutrons and/or sterile neutrons at the HIBEAM/NNBAR experiment at the European Spallation Source. *J. Phys. G*, 48(7):070501, 2021.

[60] M. Battaglieri et al. Dark matter search in a Beam-Dump eXperiment (BDX) at Jefferson Lab: an update on PR12-16-001. 12 2017.

[61] Brian Batell, Rouven Essig, and Ze'ev Surujon. Strong Constraints on Sub-GeV Dark Sectors from SLAC Beam Dump E137. *Phys. Rev. Lett.*, 113(17):171802, 2014.

[62] C. Ahdida et al. Sensitivity of the SHiP experiment to light dark matter. *JHEP*, 04:199, 2021.

[63] D. Akimov et al. Sensitivity of the COHERENT Experiment to Accelerator-Produced Dark Matter. *Phys. Rev. D*, 102(5):052007, 2020.

[64] A. A. Aguilar-Arevalo et al. Dark Matter Search in Nucleon, Pion, and Electron Channels from a Proton Beam Dump with MiniBooNE. *Phys. Rev. D*, 98(11):112004, 2018.

[65] Eder Izaguirre, Gordan Krnjaic, Philip Schuster, and Natalia Toro. New Electron Beam-Dump Experiments to Search for MeV to few-GeV Dark Matter. *Phys. Rev. D*, 88:114015, 2013.

[66] Asher Berlin, Nikita Blinov, Gordan Krnjaic, Philip Schuster, and Natalia Toro. Dark Matter, Millicharges, Axion and Scalar Particles, Gauge Bosons, and Other New Physics with LDMX. *Phys. Rev. D*, 99(7):075001, 2019.

[67] Torsten Åkesson et al. Light Dark Matter eXperiment (LDMX). 8 2018.

[68] G. Krnjaic and N. Toro. Dark matter production at intensity frontier experiments. Snowmass RF6 summary white paper, to appear on arxiv.

[69] Marco Fabbrichesi, Emidio Gabrielli, and Gaia Lanfranchi. *The Physics of the Dark Photon*. Springer International Publishing, 2021.

[70] Torsten Åkesson et al. Current Status and Future Prospects for the Light Dark Matter eXperiment. In *2022 Snowmass Summer Study*, 3 2022.

[71] Ann E. Nelson and Nikolaos Tetradis. Constraints on a New Vector Boson Coupled to Baryons. *Phys. Lett. B*, 221:80–84, 1989.

[72] Sean Tulin. New weakly-coupled forces hidden in low-energy QCD. *Phys. Rev. D*, 89(11):114008, 2014.

[73] Jeffrey M. Berryman and Susan Gardner. Neutron star structure with a new force between quarks. *Phys. Rev. C*, 104(4):045802, 2021.

[74] Alessandro Strumia. Dark Matter interpretation of the neutron decay anomaly. *JHEP*, 02:067, 2022.

[75] Bartosz Fornal and Benjamin Grinstein. Dark Matter Interpretation of the Neutron Decay Anomaly. *Phys. Rev. Lett.*, 120(19):191801, 2018. [Erratum: Phys.Rev.Lett. 124, 219901 (2020)].

[76] T. Araki et al. Search for the invisible decay of neutrons with KamLAND. *Phys. Rev. Lett.*, 96:101802, 2006.

[77] Stephen Benson, Bogdan Wojtsekhowski, Barislav Vlahovic, and Serkan Golge. Opportunities and Challenges of a Low-energy Positron Source in the LERF. *AIP Conference Proceedings of International Workshop on Physics with Positrons at Jefferson Laboratory*, 1970:050004–1, 2018.

ARIEL experiments and theory

Petr Navrátil
TRIUMF, 4004 Wesbrook Mall, Vancouver, British Columbia, V6T 2A3, Canada
E-mail: navratil@triumf.ca

Abstract. I present an overview of experiments at TRIUMF ARIEL and ISAC facilities covering both the current and the future envisioned programs. I also briefly review theory program at TRIUMF that relates to the ARIEL experimental program. I highlight several recent experimental results from the nuclear astrophysics, nuclear structure, fundamental symmetries, and the sterile neutrino search. Finally, I mention ongoing theoretical *ab initio* calculations of the proton capture on ^{7}Li related to the X17 boson observation.

1. Introduction

TRIUMF is Canada's particle accelerator centre established in 1968 as the TRI-University Meson Factory. Since 1974, TRIUMF operates the 500 MeV cyclotron accelerating the H^- ions. The high-intensity 500 MeV proton beam drives the radioactive ion beam (RIB) Isotope Separator and ACcelerator (ISAC) facility [1]. The ISAC started its operation in 1995. TRIUMF has recently embarked on the construction of ARIEL, the Advanced Rare Isotope Laboratory, with the goal to significantly expand the RIB program for Nuclear Physics, Life Sciences, and Materials Science [2]. The nuclear physics program will expand its main pillars, the nuclear structure, astrophysics, and tests of fundamental symmetries. At the heart of ARIEL there is a 100 kW, 30 MeV electron accelerator (e-linac) for isotope production via photo-production and photo-fission as well as a second proton beam line from TRIUMF's 500 MeV cyclotron for isotope production via proton-induced spallation, fragmentation, and fission. Also included in ARIEL are two production targets and related infrastructure, mass-separators and ion beam transport to ISAC, and a electron-beam ion source (EBIS) for charge breeding. Fig. 1 shows a site view of the TRIUMF accelerator complex. ARIEL will establish a multi-user capability with up to three simultaneous RIBs with more and new isotopes for TRIUMF users. The project completion is planned in 2026 with phased implementation, interleaving science with construction.

2. ISAC/ARIEL experimental stations

The short-lived isotopes produced at ISAC and ARIEL targets can be delivered to experiments using a low-energy (< 60 keV) electrostatic beam transport system or re-accelerated beams. For re-accelerated beams two acceleration stages are available, the first for energies of 0.15 to 1.8 AMeV, designed for high intensity RIBs for studies of nuclear reactions relevant for explosive nucleosynthesis, and a second superconducting stage for energies of at least 6 AMeV for masses of $A < 150$.

A wide range of experimental stations is available at the ISAC/ARIEL facility to do research in nuclear astrophysics, nuclear structure and reactions, and electro-weak interactions, see Fig. 2.

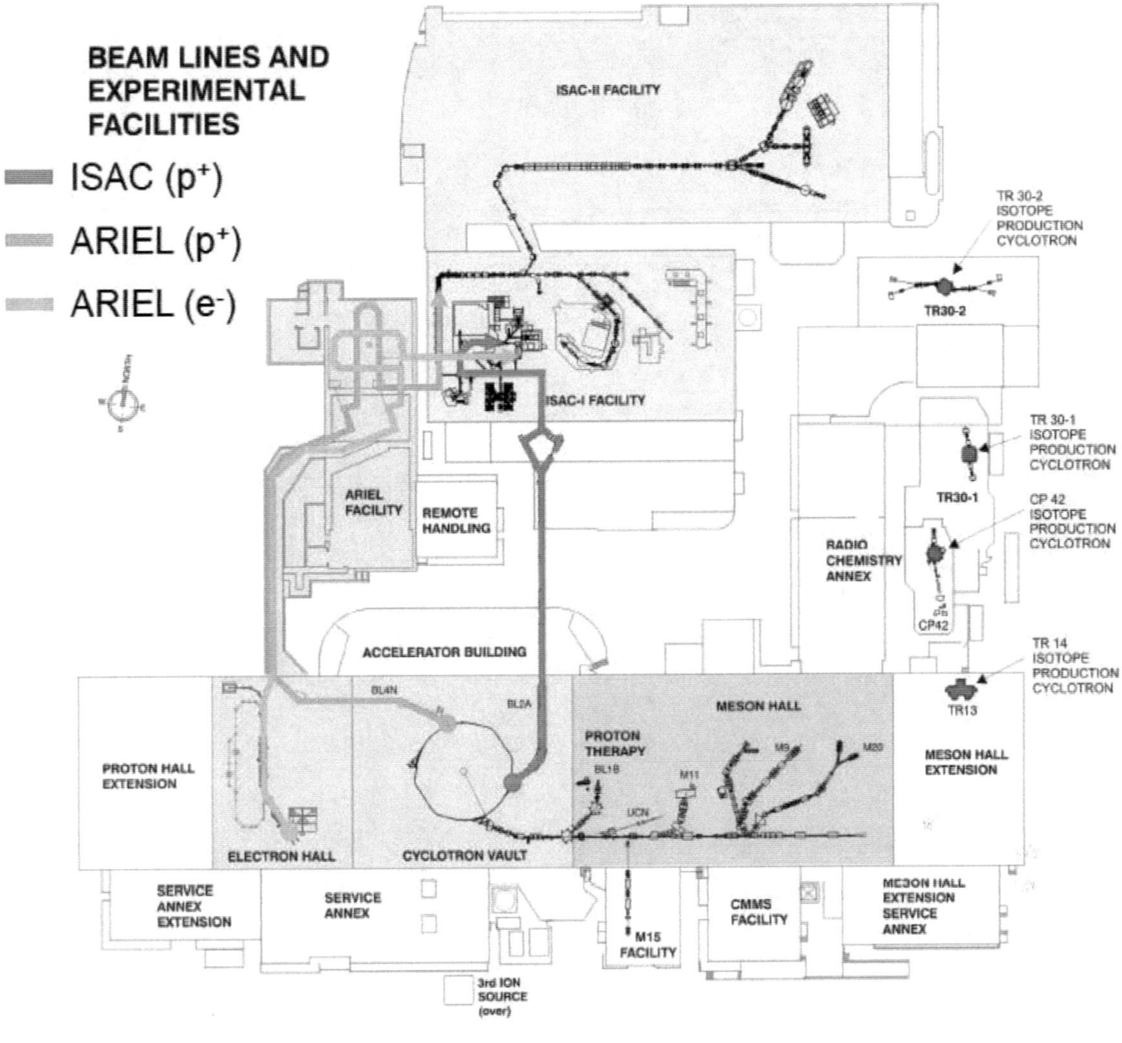

Figure 1. Site view of the TRIUMF accelerator complex.

In addition to this main focus of the facility, ISAC also hosts world-wide unique $\beta-$NMR and $\beta-$QMR set-ups for the study of magnetic properties in thin materials and at material interfaces.

Instruments focused on nuclear astrophysics research include the DRAGON (Detector of Recoils And Gamma rays Of Nuclear reactions) recoil separator designed to measure radiative capture of RIBs on hydrogen and helium target nuclei, and the TUDA (TRIUMF UK Detector Array) charged particle detector array.

The nuclear structure focused gamma-ray spectroscopy program is centred on the TRIUMF-ISAC γ-ray escape suppressed spectrometer (TIGRESS), a next generation array of high-efficiency segmented HPGe detectors with digital signal processing that is specifically designed to meet the challenges of experiments with reaccelerated RIBs. A number of auxiliary detectors are used with TIGRESS including a silicon barrel for detecting charged particles SHARC, an array of neutron detectors DESCANT and a recoil mass spectrometer EMMA.

Gamma-Ray Infrastructure For Fundamental Investigations of Nuclei (GRIFFIN) is a state-

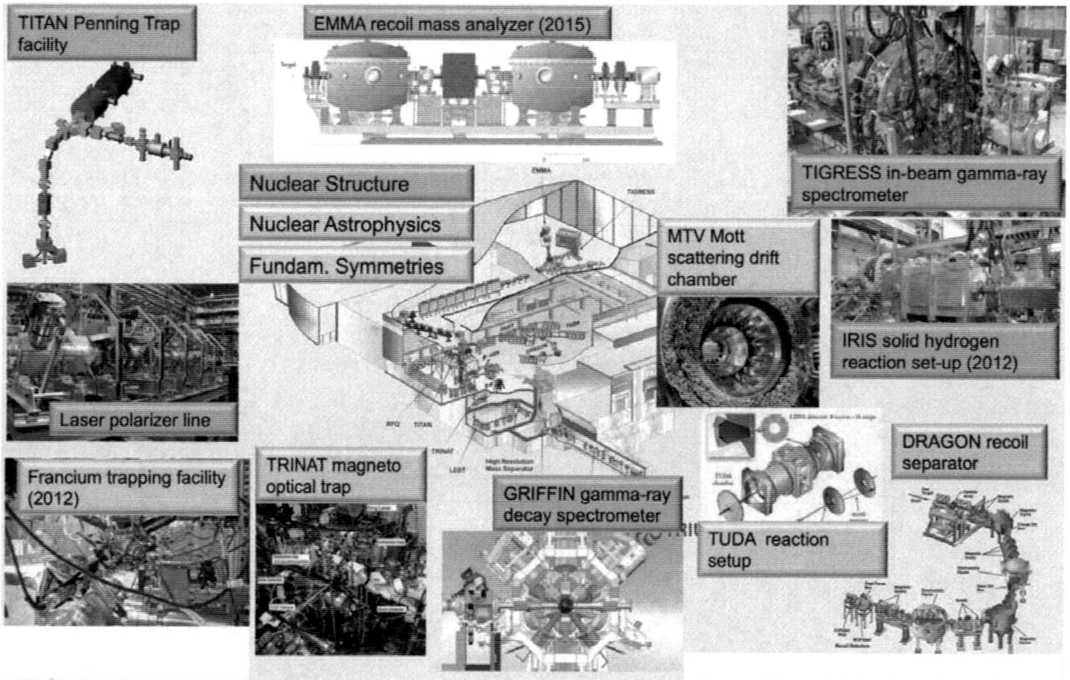

Figure 2. TRIUMF ISAC/ARIEL experimental stations. Figure courtesy of Jens Dilling.

of-the-art new high-efficiency gamma-ray spectrometer for decay spectroscopy research with the low-energy RIBs. GRIFFIN was installed in 2015 and supports a broad program in nuclear structure, nuclear astrophysics, and fundamental particle interactions.

Nuclear structure and reactions studies of exotic nuclei are also performed with the IRIS (Innovative Rare Isotope reaction Spectroscopy) facility with a frozen (solid) windowless hydrogen or deuterium target and charged particle spectrometer.

A program of high precision mass measurements is carried out at the TITAN (TRIUMF's Ion Trap for Atomic and Nuclear science) Penning trap facility. Research focus includes nuclear structure studies of exotic halo nuclei [3], shell evolution in neutron rich isotopes [4, 5, 6], precise determination of Q-values of superallowed β-decays used for the determination of the V_{ud} matrix element of the CKM matrix [7].

Precision measurements of the angular correlations between electron (positron) and anti-neutrino (neutrino) emitted in the β-decay of spin aligned nuclei allows to search for new components of the electro-weak interaction, such as the exchange of new bosons not described in the Standard Model. Since the neutrino cannot be detected the correlation is determined from the measurement of the momenta of the electron and recoiling nucleus. The TRINAT experiment employs and magneto-optical trap (MOT) for such experiments. Recently, the TRINAT collaboration measured the β asymmetry with respect to the initial nuclear spin in ^{37}K that provided new constraints on physics beyond the standard model [8].

The Francium trapping facility aims at measurements of the nuclear anapole moment in unstable Fr isotopes that provides a unique access to neutral weak interactions inside the nucleus [9].

3. Theory program at TRIUMF

TRIUMF maintains an active theory program focused on *ab initio* nuclear theory, nuclear astrophysics, and particle physics. The nuclear theory research is well aligned with the ISAC/ARIEL experiments.

First principles or *ab initio* nuclear theory describes nuclei as systems of protons and neutrons interacting by nucleon-nucleon (NN) and three-nucleon (3N) interactions derived from the Quantum Chromodynamics by means of the chiral effective field theory (EFT). With these interactions as input, one solves many-nucleon Schrödinger equation to predict properties of atomic nuclei. Unique to TRIUMF nuclear theory is the development of a unified approach to nuclear structure and reactions for light nuclei, the No-Core Shell Model with Continuum (NCSMC) method [10], and, further, a powerful valence-space method for medium mass nuclei, the Valence-Space In-Medium Similarity Renormalization Group (VS-IMSRG) method [11, 12]. These approaches involve large-scale high-performance computing using massively parallel codes running on supercomputers including Summit at ORNL and Niagara of the Digital Research Alliance of Canada. Applications of *ab initio* nuclear theory include studies of the structure exotic nuclei, nuclei far from stability, nuclear reactions important for astrophysics, tests of fundamental symmetries and searches for physics beyond the standard model.

Nuclear astrophysics efforts center on theoretical investigations and modelling of the r-process nucleosynthesis. ARIEL will produce unprecedented intensities of rare isotope beams of neutron-rich nuclei through photo-fission that are synthesized in the r-process in the Cosmos. In synergy with the with precision astronomy and sophisticated astrophysical modelling, studies of the neutron rich nuclei will allow us to identify the astrophysical site of the r-process.

Particle theory at TRIUMF focuses on dark matter physics, collider phenomenology, neutrino physics, particle cosmology, and hadronic physics.

4. Recent research highlights

This section presents a selection of recent results obtained by ISAC/ARIEL experiments. Synergy with the TRIUMF theory is highlighted.

4.1. Mass measurements of neutron-rich Titanium and Fe isotopes

TITAN collaboration performed a precision mass investigation of the neutron-rich titanium isotopes $^{51-55}$Ti [4]. The range of the measurements covered the N=32 shell closure, and the overall uncertainties of the $^{52-55}$Ti mass values were significantly reduced. Obtained results conclusively established the existence of the weak shell effect at N=32, narrowing down the abrupt onset of this shell closure. Experimental data were compared with state-of-the-art *ab initio* calculations including the in-house VS-IMSRG. In addition, a locally developed chiral EFT 3N interaction was applied and tested for the first time. These calculations, despite very successfully describing where the N=32 shell gap is strong, overpredicted its strength and extent in titanium and heavier isotones.

Still more recently, high-precision mass measurements of neutron-rich Fe isotopes were performed at the TITAN facility using the newly commissioned multiple-reflection time-of-flight mass spectrometer (MR-ToF-MS) [5]. These measurements were accompanied by in-house *ab initio* VS-IMSRG calculations which enabled theoretical assignment of the spin-parities of the ^{69}Fe ground and isomeric states. Together with mean-field calculations of quadrupole deformation parameters for the Fe isotope chain, these results benchmarked a maximum of deformation in the N=40 island of inversion in Fe.

4.2. ^{10}C elastic scattering with protons and nuclear forces

TRIUMF IRIS reaction spectroscopy station was used to measure elastic scattering of the exotic ^{10}C nucleus on protons in inverse kinematics [13]. The ^{10}C nucleus is located at the proton drip

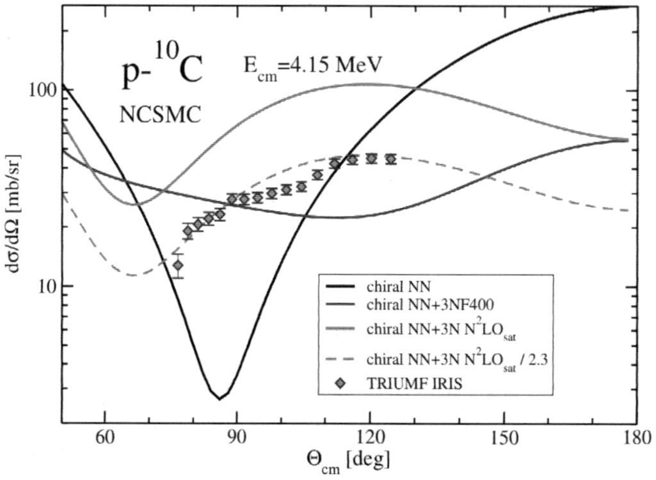

Figure 3. Measured differential cross section for ^{10}C$(p,p)^{10}$C$_{gs}$ at $E_{c.m.}$=4.15 MeV. The curves are *ab initio* theory calculations. The black, blue, and red solid curves are with the chiral NN, NN+3N400, N^{2}LO$_{\rm sat}$ interactions, respectively. The red dashed curve is the N^{2}LO$_{\rm sat}$ calculation scaled down by a factor of 2.3. Further details can be found in Ref. [13].

line with its isotonic neighbours ^{9}B, ^{8}Be, and ^{11}N unbound. The chosen energies covered a low-energy resonance region in the composite unbound ^{11}N nucleus, a mirror of the halo nucleus ^{11}Be famous for its parity-inverted ground state. The measured differential cross section was compared to *ab initio* NCSMC calculations with different chiral EFT interactions as input, see Fig. 3. The shape of the cross section was reproduced only with the chiral NN+3N interaction that simultaneously reproduces the ^{11}Be parity inversion. The experiment demonstrated a strong sensitivity of the elastic scattering off the extremely exotic ^{10}C to the nuclear force prescription.

4.3. Coulomb-excitation measurements of ^{23}Mg and ^{23}Na using the TIGRESS spectrometer
Coulomb-excitation measurements of the $|T_z|$=1/2 mirror nuclei ^{23}Mg and ^{23}Na, where the former is unstable and the latter stable, were performed at the TRIUMF-ISAC facility using the TIGRESS spectrometer [14] . They were used to determine the $E2$ matrix elements of mixed $E2/M1$ transitions between the ground and the first excited state with the goal to challenge *ab initio* VS-IMSRG nuclear theory. The experiment improved the precision of the extracted $E2$ strength by a factor of six and demonstrated an underprediction of the strength by the *ab initio* theory. This points to the need of including higher-order operator terms to improve the theoretical description.

4.4. Radioactive capture on nuclear isomers
The presence of radioactive ^{26g}Al ($t_{1/2}$ = 7.2 × 10^5 yr) in the universe has been inferred from the detection of characteristic 1.809-MeV γ rays throughout the interstellar medium and observations of ^{26}Mg excesses in meteorites. These discoveries provided direct evidence for ongoing nucleosynthesis in our Galaxy. However, stellar nucleosynthesis of ^{26}Al is complicated by the presence of a 0^+ isomer, ^{26m}Al ($t_{1/2}$ = 6.3 s) and, in particular, its destruction by a proton capture reaction needs to be understood. TRIUMF DRAGON spectrometer was used to perform a direct measurement of the ^{26m}Al(p,γ) reaction [15]. This is the first direct measurement of an astrophysical reaction using a radioactive beam of isomeric nuclei. The isomeric component of

the incoming beam was identified by its associated super-allowed β^+ decay to the ^{26}Mg ground state. The strength of a key resonance has been determined which enabled to establish the reaction rate in astrophysical relevant temperature range.

4.5. The BeEST experiment

Sterile neutrinos are natural extensions to the standard model of particle physics. A new search for the existence of sub-MeV sterile neutrinos using the decay-momentum reconstruction technique in the decay of ^{7}Be has been established by a Colorado School of Mines-TRIUMF-LLNL Beryllium Electron capture in Superconducting Tunnel junctions (BeEST) collaboration [16, 17, 18]. The experiment measures the total energy of the ^{7}Li daughter atom from the electron capture decay of ^{7}Be implanted into sensitive superconducting tunnel junction (STJ) quantum sensors. The ^{7}Be production and implantation is executed at the TRIUMF ISAC/ARIEL facility. Despite its recent launch, the experiment already significantly improved the limits on the existence of sterile neutrinos in the 100–850 keV mass range [17]. TRIUMF particle theorists contribute to the BeEST program.

5. Future experimental facilities at ARIEL

Several new facilities as well as enhancements of existing experiments are envisioned.

5.1. Delivery of spin-polarized beams to GRIFFIN

It is planned to deliver spin-polarized beams to the GRIFFIN high-efficiency gamma-ray spectrometer. A new 2-meter section of beamline is needed to join the Polarizer beamline to the GRIFFIN spectrometer. This will allow to perform decay spectroscopy with $\beta - \gamma$ and $\gamma - \gamma$ coincidences of spin-polarized beams including γ-tagged β asymmetry for firm assignment of spins, parities; isospin mixing measurements relevant to the V_{ud} determination; searches for time-reversal breaking. Further, decay spectroscopy of isomerically pure beams by resonant photoionization will become feasible.

5.2. Radioactive Molecule (RadMol) Laboratory

Building on TRIUMF's capability to produce a large variety of radioactive ion beams, the international RadMol collaboration seeks to establish a dedicated laboratory for radioactive molecules and fundamental physics at TRIUMF. It will host three independent experimental stations which are directly coupled via a low-energy beamline to TRIUMF's RIB facilities ARIEL and ISAC. The laboratory will also provide space and infrastructure for the ion reaction cell as one formation site for radioactive molecules. The facility plans include a dedicated laser laboratory.

Radioactive molecules can be used as novel precision probes for fundamental physics. Initial physics program will study octuple-deformed nuclei incorporated into polar molecules which provide unmatched sensitivity for nuclear electric dipole moment (EDM). Utilization of diatomic molecules will enable access nuclear anapole moments. Expansions into other fields are also envisioned by the RadMol collaboration.

5.3. TRIUMF Storage Ring (TRISR)

A feasibility study is currently underway for a new worldwide unique facility, the TRIUMF Storage Ring (TRISR) Project for the direct measurement of neutron capture cross sections on short-lived radioactive nuclei down to seconds half-lives in inverse kinematics. The TRISR will be coupled to the ISAC/ARIEL radioactive beam facility and include a high-flux neutron generator producing a "neutron target" that will intersect with the orbiting ion beam.

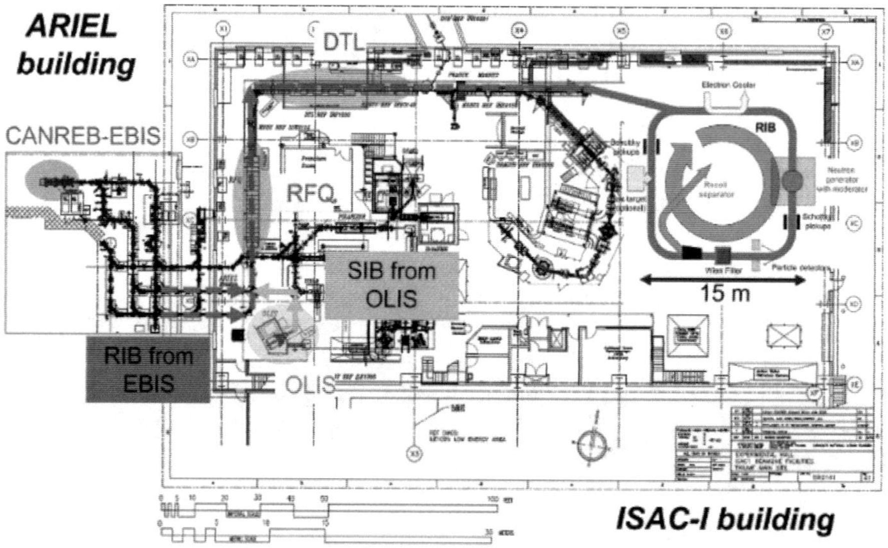

Figure 4. Proposed TRIUMF Storage Ring for a direct measurement of the neutron capture on radioactive nuclei in inverse kinematics. Figure courtesy of Iris Dillmann.

Combining the ARIEL e-linac photofission RIB production with the CANREB EBIS will enable cleaner high-intensity neutron-rich beams which are perfectly suited for direct injection into the storage ring. The ion beam orbiting in the storage ring allows multiple passes (100 kHz - few MHz frequency over several seconds) through the target area and thus provides a much higher luminosity than one-pass experiments, enabling a unique physics program. The TRISR will be a low-energy ring (E= 0.1−2 AMeV) with a 15m x 15m footprint and will be located next to the DRAGON spectrometer in the ISAC-I experimental hall, see Fig. 4.

The neutron generator and moderator design is presently under discussion. A compact-size high-flux DT neutron generator from the US-based company SHINE Technologies, LLC with an unmoderated neutron flux up to 5×10^{13} n/s is the favoured candidate. The physics program of the TRISR will center on nuclear astrophysics applications, in particular measurements of radiative neutron capture reaction cross sections on unstable nuclei with A>50 required for a reliable modelling of the intermediate (i-) and rapid (r-) neutron capture process nucleosynthesis. When no radioactive beam is available, the installation of a diagonal neutron activation position at the neutron generator will in addition allow the production of short-lived radioactive nuclei for use in local radiopharmaceutical research projects (e.g., ^{161}Tb and ^{177}Lu for Targeted Radionuclide Therapy and SPECT imaging), as well as for material and environmental science projects to detect trace amounts of stable isotopes via Instrumental Neutron Activation Analysis (INAA).

6. Theoretical calculations related to the X17 boson

The DarkLight experiment, a magnetic spectrometer system, will be installed in the ARIEL e-linac beam hall in the next year, with an initial program to search for corroborating evidence of the new physics proposed to explain current anomalies in ^{8}Be and ^{4}He decay. These anomalies observed in the electron-positron pair production from the proton capture on ^{7}Li and ^{3}H were reported by the ATOMKI collaboration and interpreted by the decay of a new boson called X17 with the mass ∼17 MeV [19, 20, 21, 22]. The latest developments in the ATOMKI experiments

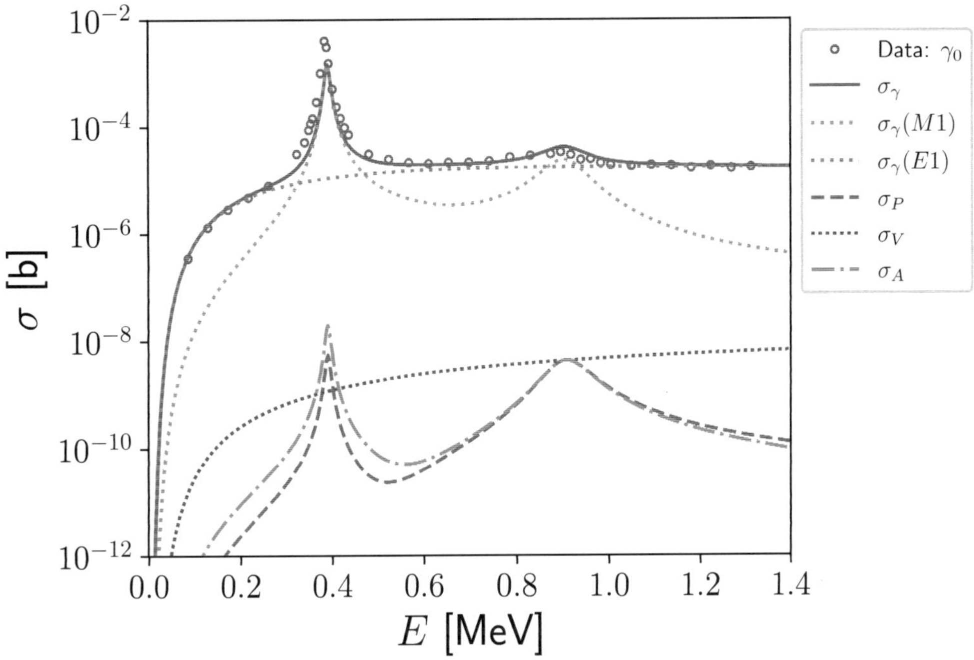

Figure 5. The proton capture cross section on ^{7}Li dependence on the center-of-mass energy. The calculated γ emission cross section is compared to experimental data from Ref. [23] and to the calculated hypothetical X17 boson emissions considering the $E1$ vector, pseudoscalar (axion), and axial vector boson candidates. All transitions are to the ^{8}Be 0^+ ground state. Calculations are preliminary, details will be given in Ref. [24].

and the progress of the DarkLight installation were reported at the workshop New Scientific Opportunities at the TRIUMF ARIEL e-linac.

TRIUMF theorists were active in addressing the X17 boson anomaly shortly after it was initially reported [25]. More recently, we embarked on a more in-depth approach to the $p+^7$Li capture reaction within the *ab initio* NCSMC approach [10]. This way one can describe simultaneously the structure of ^{8}Be, the elastic and inelastic ^{7}Li(p,p) scattering, the charge exchange reaction ^{7}Li$(p,n)^7$Be, the γ capture ^{7}Li$(p,\gamma)^8$Be, the pair production ^{7}Li$(p,e^+e^-)^8$Be as well as the X17 boson production ^{7}Li$(p,X)^8$Be and decay for a variety of candidates for the hypothetical boson [24]. The input for these calculations are chiral EFT NN+3N interactions. Preliminary NCSMC cross section results are shown in Fig. 5. The calculated ^{7}Li$(p,\gamma)^8$Be radiative capture cross section compares well with the data from Ref. [23]. The electric dipole ($E1$) and the magnetic dipole ($M1$) contributions are shown separately. The two peaks dominated by the $M1$ contributions are due to the two 1^+ resonances in ^{8}Be with the lower one predominantly isospin T=1 while the higher one predominantly T=0 with a suppressed $M1$ decay to the ^{8}Be 0^+ T=0 ground state. Cross sections for the emission of the X17 boson lower by several orders of magnitude are also shown. Three X17 candidates were considered, a pseudo scalar (axion), axial vector, and $E1$ vector using operators from Refs. [25, 26, 27, 28]. The electron-positron pair production cross section (not shown in the figure) has a shape basically identical to the γ capture cross section with its magnitude scaled by a factor of $\sim \alpha/2\pi \sim 10^{-3}$.

One could understand then that no anomaly would be observed in the first resonance with the very high electromagnetic $M1$ rate. An effect from the hypothetical boson can be expected in the second 1^+ resonance where both the pseudoscalar and the axial vector boson candidate cross sections peak. An anomaly between the 1^+ resonances would be consistent with the $E1$ vector. The latter is the preferred candidate according to the latest ATOMKI publications [21, 22].

7. Conclusions

With ARIEL completion within the next five years, TRIUMF will achieve a multi-user capability with up to three simultaneous RIBs. This will allow an expansion of the Nuclear Physics, Life Sciences, and Materials Science program. In synergy with theory, nuclear physics experimental program will be positioned to enhance its science output in all three main pillars, the nuclear structure, astrophysics, and tests of fundamental symmetries. At the same time, ARIEL e-linac will enable particle physics experiments such as DarkLight.

Acknowledgments

I would like to thank Iris Dillmann for useful comments. This work was supported by the NSERC Grant No. SAPIN-2022-00019. TRIUMF receives federal funding via a contribution agreement with the National Research Council of Canada. Computing support came from an INCITE Award on the Summit supercomputer of the Oak Ridge Leadership Computing Facility (OLCF) at ORNL, from Livermore Computing, and the Digital Research Alliance of Canada.

References

[1] Dilling J, Krücken R and Ball G 2014 *Hyperfine Interactions* **225** 1–8 URL `https://doi.org/10.1007/s10751-013-0877-7`

[2] Dilling J, Krücken R and Merminga L 2014 *Hyperfine Interactions* **225** 253–262 URL `https://doi.org/10.1007/s10751-013-0906-6`

[3] Brodeur M, Brunner T, Champagne C, Ettenauer S, Smith M J, Lapierre A, Ringle R, Ryjkov V L, Bacca S, Delheij P, Drake G W F, Lunney D, Schwenk A and Dilling J 2012 *Phys. Rev. Lett.* **108**(5) 052504 URL `https://link.aps.org/doi/10.1103/PhysRevLett.108.052504`

[4] Leistenschneider E, Reiter M P, Ayet San Andrés S, Kootte B, Holt J D, Navrátil P, Babcock C, Barbieri C, Barquest B R, Bergmann J, Bollig J, Brunner T, Dunling E, Finlay A, Geissel H, Graham L, Greiner F, Hergert H, Hornung C, Jesch C, Klawitter R, Lan Y, Lascar D, Leach K G, Lippert W, McKay J E, Paul S F, Schwenk A, Short D, Simonis J, Somà V, Steinbrügge R, Stroberg S R, Thompson R, Wieser M E, Will C, Yavor M, Andreoiu C, Dickel T, Dillmann I, Gwinner G, Plaß W R, Scheidenberger C, Kwiatkowski A A and Dilling J 2018 *Phys. Rev. Lett.* **120**(6) 062503 URL `https://link.aps.org/doi/10.1103/PhysRevLett.120.062503`

[5] Porter W S, Ashrafkhani B, Bergmann J, Brown C, Brunner T, Cardona J D, Curien D, Dedes I, Dickel T, Dudek J, Dunling E, Gwinner G, Hockenbery Z, Holt J D, Hornung C, Izzo C, Jacobs A, Javaji A, Kootte B, Kripkó-Koncz G, Lykiardopoulou E M, Miyagi T, Mukul I, Murböck T, Plaß W R, Reiter M P, Ringuette J, Scheidenberger C, Silwal R, Walls C, Wang H L, Wang Y, Yang J, Dilling J and Kwiatkowski A A 2022 *Phys. Rev. C* **105**(4) L041301 URL `https://link.aps.org/doi/10.1103/PhysRevC.105.L041301`

[6] Silwal R, Andreoiu C, Ashrafkhani B, Bergmann J, Brunner T, Cardona J, Dietrich K, Dunling E, Gwinner G, Hockenbery Z, Holt J, Izzo C, Jacobs A, Javaji A, Kootte B, Lan Y, Lunney D, Lykiardopoulou E, Miyagi T, Mougeot M, Mukul I, Murböck T, Porter W, Reiter M, Ringuette J, Dilling J and Kwiatkowski A 2022 *Physics Letters B* **833** 137288 ISSN 0370-2693 URL `https://www.sciencedirect.com/science/article/pii/S0370269322004221`

[7] Reiter M P, Leach K G, Drozdowski O M, Stroberg S R, Holt J D, Andreoiu C, Babcock C, Barquest B, Brodeur M, Finlay A, Foster M, Gallant A T, Gwinner G, Klawitter R, Kootte B, Kwiatkowski A A, Lan Y, Lascar D, Leistenschneider E, Lennarz A, Paul S, Steinbrügge R, Thompson R I, Wieser M and Dilling J 2017 *Phys. Rev. C* **96**(5) 052501 URL `https://link.aps.org/doi/10.1103/PhysRevC.96.052501`

[8] Fenker B, Gorelov A, Melconian D, Behr J A, Anholm M, Ashery D, Behling R S, Cohen I, Craiciu I, Gwinner G, McNeil J, Mehlman M, Olchanski K, Shidling P D, Smale S and Warner C L 2018 *Phys. Rev. Lett.* **120**(6) 062502 URL `https://link.aps.org/doi/10.1103/PhysRevLett.120.062502`

[9] Kalita M R, Behr J A, Gorelov A, Pearson M R, DeHart A C, Gwinner G, Kossin M J, Orozco L A, Aubin S, Gomez E, Safronova M S, Dzuba V A and Flambaum V V 2018 *Phys. Rev. A* **97**(4) 042507 URL `https://link.aps.org/doi/10.1103/PhysRevA.97.042507`

[10] Navrátil P, Quaglioni S, Hupin G, Romero-Redondo C and Calci A 2016 *Physica Scripta* **91** 053002 URL `https://doi.org/10.1088/0031-8949/91/5/053002`

[11] Tsukiyama K, Bogner S K and Schwenk A 2011 *Phys. Rev. Lett.* **106**(22) 222502 URL `http://link.aps.org/doi/10.1103/PhysRevLett.106.222502`

[12] Bogner S K, Hergert H, Holt J D, Schwenk A, Binder S, Calci A, Langhammer J and Roth R 2014 *Phys. Rev. Lett.* **113**(14) 142501 URL `http://link.aps.org/doi/10.1103/PhysRevLett.113.142501`

[13] Kumar A, Kanungo R, Calci A, Navrátil P, Sanetullaev A, Alcorta M, Bildstein V, Christian G, Davids B, Dohet-Eraly J, Fallis J, Gallant A T, Hackman G, Hadinia B, Hupin G, Ishimoto S, Krücken R, Laffoley A T, Lighthall J, Miller D, Quaglioni S, Randhawa J S, Rand E T, Rojas A, Roth R, Shotter A, Tanaka J, Tanihata I and Unsworth C 2017 *Phys. Rev. Lett.* **118**(26) 262502 URL `https://link.aps.org/doi/10.1103/PhysRevLett.118.262502`

[14] Henderson J, Hackman G, Ruotsalainen P, Holt J D, Stroberg S R, Andreoiu C, Ball G C, Bernier N, Bowry M, Caballero-Folch R, Cruz S, Diaz Varela A, Evitts L J, Frederick R, Garnsworthy A B, Holl M, Lassen J, Measures J, Olaizola B, O'Sullivan E, Paetkau O, Park J, Smallcombe J, Svensson C E, Whitmore K and Wu C Y 2022 *Phys. Rev. C* **105**(3) 034332 URL `https://link.aps.org/doi/10.1103/PhysRevC.105.034332`

[15] Lotay G, Lennarz A, Ruiz C, Akers C, Chen A A, Christian G, Connolly D, Davids B, Davinson T, Fallis J, Hutcheon D A, Machule P, Martin L, Mountford D J and Murphy A S J 2022 *Phys. Rev. Lett.* **128**(4) 042701 URL `https://link.aps.org/doi/10.1103/PhysRevLett.128.042701`

[16] Fretwell S, Leach K G, Bray C, Kim G B, Dilling J, Lennarz A, Mougeot X, Ponce F, Ruiz C, Stackhouse J and Friedrich S 2020 *Phys. Rev. Lett.* **125**(3) 032701 URL `https://link.aps.org/doi/10.1103/PhysRevLett.125.032701`

[17] Friedrich S, Kim G B, Bray C, Cantor R, Dilling J, Fretwell S, Hall J A, Lennarz A, Lordi V, Machule P, McKeen D, Mougeot X, Ponce F, Ruiz C, Samanta A, Warburton W K and Leach K G 2021 *Phys. Rev. Lett.* **126**(2) 021803 URL `https://link.aps.org/doi/10.1103/PhysRevLett.126.021803`

[18] Leach K G and Friedrich S 2021 The beest experiment: Searching for beyond standard model neutrinos using ^{7}be decay in stjs URL `https://arxiv.org/abs/2112.02029`

[19] Krasznahorkay A J, Csatlós M, Csige L, Gácsi Z, Gulyás J, Hunyadi M, Kuti I, Nyakó B M, Stuhl L, Timár J, Tornyi T G, Vajta Z, Ketel T J and Krasznahorkay A 2016 *Phys. Rev. Lett.* **116**(4) 042501 URL `https://link.aps.org/doi/10.1103/PhysRevLett.116.042501`

[20] Firak, D S, Krasznahorkay, A J, Csatlós, M, Csige, L, Gulyás, J, Koszta, M, Szihalmi, B, Timár, J, Nagy, Á, Sas, N J and Krasznahorkay, A 2020 *EPJ Web Conf.* **232** 04005 URL `https://doi.org/10.1051/epjconf/202023204005`

[21] Krasznahorkay A J, Csatlós M, Csige L, Gulyás J, Krasznahorkay A, Nyakó B M, Rajta I, Timár J, Vajda I and Sas N J 2021 *Phys. Rev. C* **104**(4) 044003 URL `https://link.aps.org/doi/10.1103/PhysRevC.104.044003`

[22] Sas N J, Krasznahorkay A J, Csatlós M, Gulyás J, Kertész B, Krasznahorkay A, Molnár J, Rajta I, Timár J, Vajda I and Harakeh M N 2022 Observation of the x17 anomaly in the ^{7}li(p,e^+e^-)^{8}be direct proton-capture reaction URL `https://arxiv.org/abs/2205.07744`

[23] Zahnow D, Angulo C, Rolfs C, Schmidt S, Schulte W H and Somorjai E 1995 *Zeitschrift für Physik A Hadrons and Nuclei* **351** 229–236 URL `https://doi.org/10.1007/BF01289534`

[24] Gysbers P, Navrátil P, Kravvaris K, Hupin G and Quaglioni S 2022 in preparation

[25] Kozaczuk J, Morrissey D E and Stroberg S R 2017 *Phys. Rev. D* **95**(11) 115024 URL `https://link.aps.org/doi/10.1103/PhysRevD.95.115024`

[26] Donnelly T W, Freedman S J, Lytel R S, Peccei R D and Schwartz M 1978 *Phys. Rev. D* **18**(5) 1607–1620 URL `https://link.aps.org/doi/10.1103/PhysRevD.18.1607`

[27] Zhang X and Miller G A 2021 *Physics Letters B* **813** 136061 ISSN 0370-2693 URL `https://www.sciencedirect.com/science/article/pii/S0370269321000010`

[28] Backens J and Vanderhaeghen M 2022 *Phys. Rev. Lett.* **128**(9) 091802 URL `https://link.aps.org/doi/10.1103/PhysRevLett.128.091802`

Journal of Physics: Conference Series **2391** (2022) 012003 doi:10.1088/1742-6596/2391/1/012003

ARIEL Accelerator Overview

Thomas Planche [1], Katherine Pachal [1], Richard Milner [2]
[1] TRIUMF, Vancouver, BC, Canada
[2] Massachusetts Institute of Technology, Cambridge, MA, USA
E-mail: tplanche@triumf.ca

Abstract. This paper gives an overview of the ARIEL electron accelerator including the existing facility and possible future upgrade possibilities.

1. Overview of the Existing Facility

Fig. 1 shows a schematic layout of the TRIUMF accelerator complex, which is Canada's particle accelerator centre. Users of this unique accelerator complex include both Canadian and international collaborators. The primary beam driver is a 500 MeV H^- cyclotron, currently running at 100 kW with a near-future plan to increase to 150 kW. It is used to produce rare isotopes, neutrons and muons. Several other cyclotrons of energy 14-42 MeV are used for medical isotope production and have a total beam power of about 100 kW.

In addition, there are:

- the Isotope Separator and Accelerator (ISAC) facility;
- the Isotope Separator OnLine (ISOL) facility
- ISAC-I normal conducting linac, 0.15 - 1.8 MeV/u
- ISAC-II superconducting linac, 1.5-16.5 MeV/u .

The Advanced Rare IsotopE Laboratory (ARIEL), currently under construction, will be served by a superconducting electron continuous wave (CW) linac with an energy of 30 MeV and power capability up to 300 kW. This linac is already operational, though with average beam power output currently limited to 10 kW by the rating of the present beam dump.

2. Electron Linac: Design Parameters and Current Machine Performance

Fig. 2 shows a schematic layout of the existing ARIEL electron linac. The main components are:

Electron gun: A 300 kV thermionic gun produces bunches at 650 MHz and delivers up to 10mA average beam current in CW mode [1].

Linac: This consists of three 9-cell 1.3 GHz niobium cavities working at 2 K in two different cryomodules [2, 3, 4, 5]. The injector cryomodule contains one cavity and the second cryomodule contains two. There are two 50 kW RF couplers per cavity.

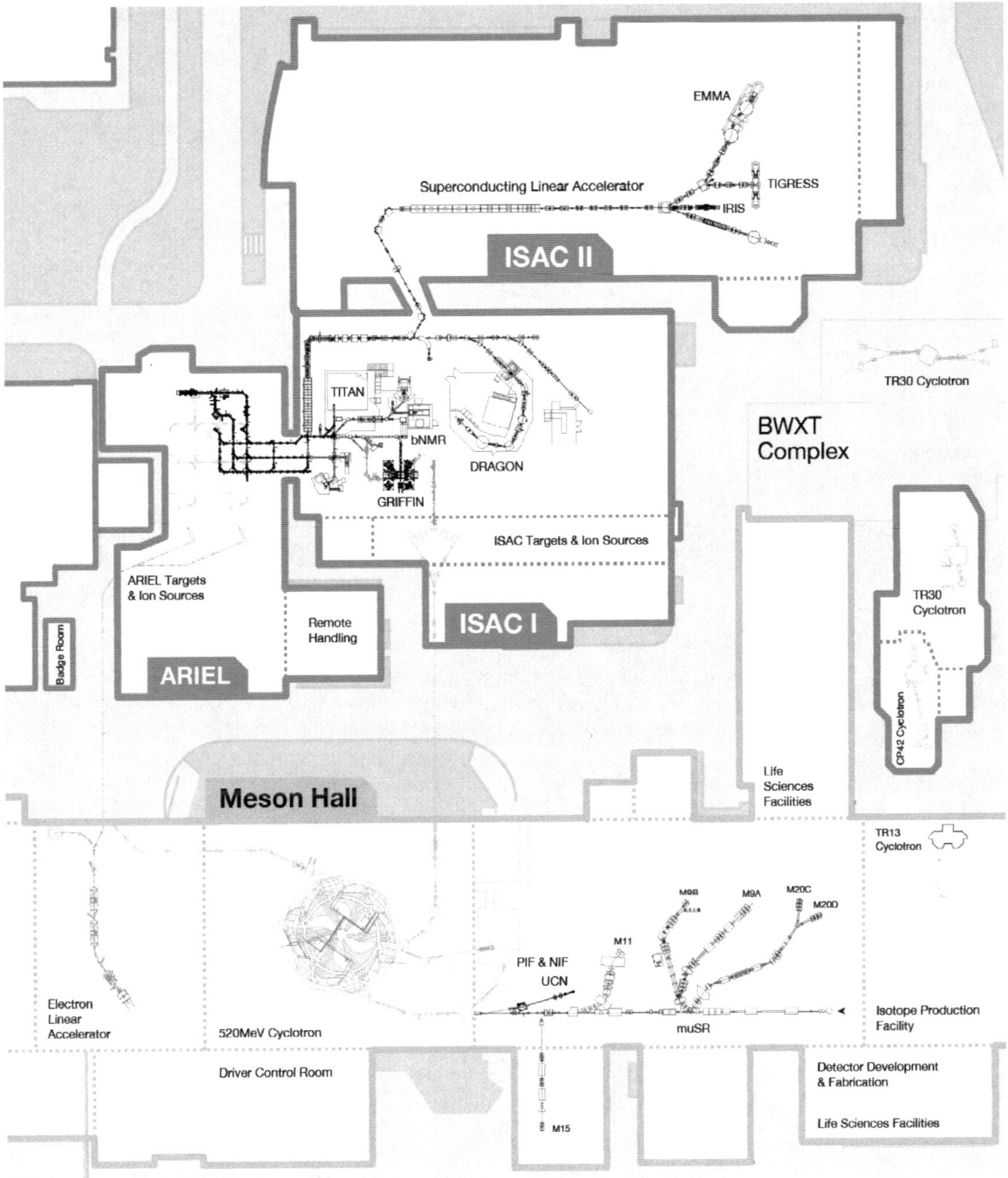

Figure 1. Schematic layout of the existing TRIUMF accelerator complex.

RF sources: There are two CPI 290 kW 1.3 GHz klystrons in the e-hall, each with a power supply positioned on the e-hall roof. A single RF source is used for the two cavities in the second cryomodule.

Cryogenics: A 800W (@4 K) helium liquefier is used. There are three 2 K Busch sub-atmospheric pumps in operation, with one 'spare'.

Beam dump: The current beam dump is rated at 10 kW.

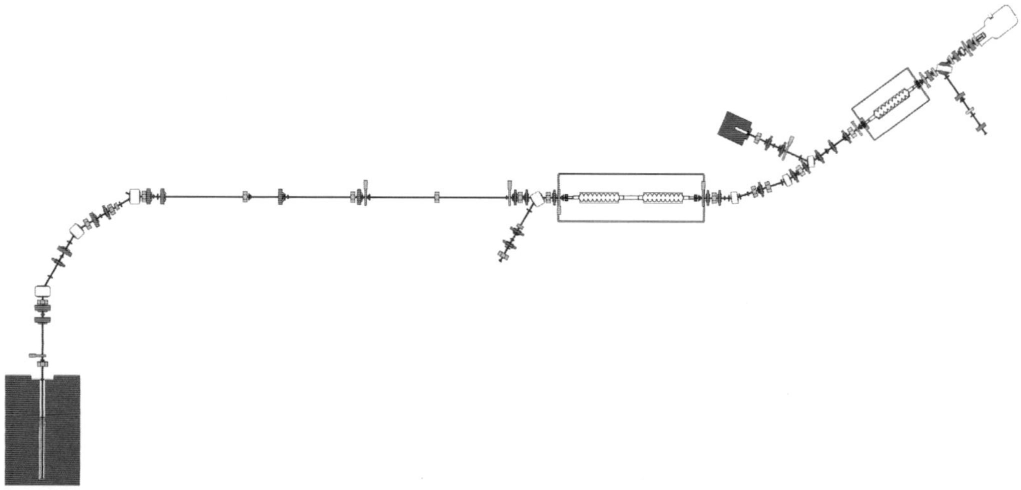

Figure 2. Schematic layout of the existing ARIEL electron linac (in red).

Since September of 2021, the e-linac has demonstrated ability to run at 30 MeV energy and 10 kW power. However, reliability of beam delivery is not yet meeting specifications. This can be seen in Figure 3, which shows the peak beam current in a typical day's running at 10 kW. The sharp drops in current correspond to beam trips and overall instability.

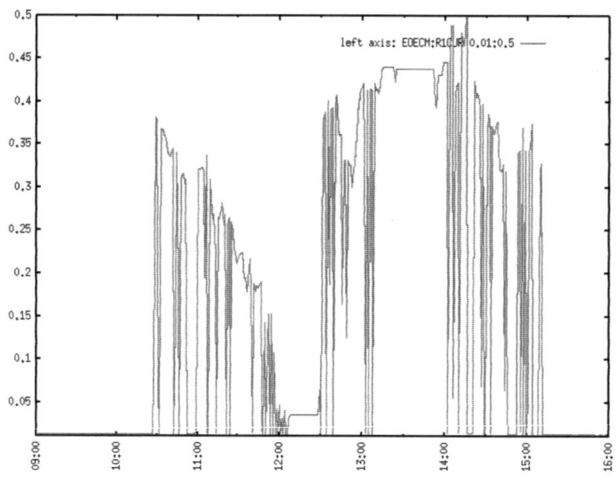

Figure 3. Example peak beam current in mA from a run at 10 kW. The fluctuations show unreliable operation.

The top four issues causing e-linac downtime are:

e-Gun: High voltage and beam current stability present ongoing challenges.

Tunability: Many procedures depend on tedious manual interventions. This also complicates the training of operators.

Journal of Physics: Conference Series **2391** (2022) 012003 doi:10.1088/1742-6596/2391/1/012003

Spurious trips: The origin of these trips is understood. They are caused by a series of hardware, interlock and diagnostics related issues, which are being addressed in our current reliability campaign.

Lack of spare parts: Simple failures can cause long delays.

The team has established strategies to address these issues over the coming year. The objective is to demonstrate three-day continuous beam delivery by March 2023.

3. Pre-ARIEL Science in the Electron Hall

The ARIEL electron target will be ready to take its first beam in 2025. Although the e-linac was built to deliver beam to the ARIEL electron target, it has the potential to become a multi-purpose, multi-user facility. With the e-Linac already capable of producing 30 MeV beam, the intervening time before ARIEL comes online enables the establishment of an early science program using the e-linac for other experiments. The following experimental programs have been established in the e-hall and are or will be completed before ARIEL begins taking beam.

ARIEL targets: A converter test stand was used to validate the target technology planned for ARIEL. The 300 keV test stand operating at powers up to 3 kW can simulate the sort of power density that will be encountered in the future 30 MeV/100 kW ARIEL targets. These tests allowed a gold-on-aluminum option to be ruled out and a tantalum-on-aluminum target to be validated as a viable option [6].

FLASH: An experiment to research radio therapy is in progress. This is a life science experiment situated at the 10 MeV beam dump and exploring high dose rate radiotherapy [7].

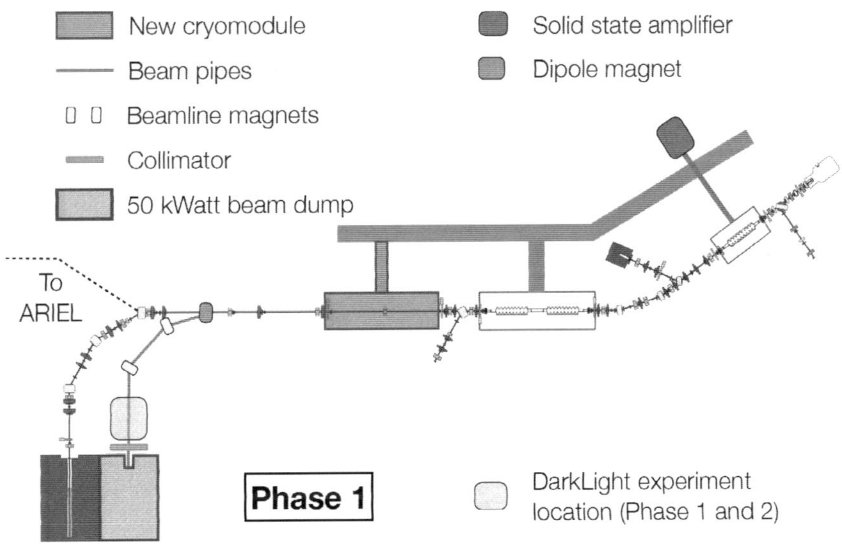

Figure 4. Schematic layout of the DarkLight experiment in Phase 1 (2024), when it will operate at 50 MeV as a single, dedicated user at the e-linac.

DarkLight: The CW, high-intensity electron driver for ARIEL is ideal for the type of thin-target scattering experiment proposed by the DarkLight collaboration. The DarkLight target will be placed close to the existing high-power dump for an initial round of data taking in the fall of 2023. The current beam energy of 30 MeV is somewhat low for the physics goals of the experiment and an energy upgrade to 50 MeV is required to probe the 17 MeV mass range of the anomaly observed by the ATOMKI experiment [8]. This energy upgrade will be achieved by the addition of another two-cavity cryomodule matching the existing second cryomodule to the beamline and is expected to complete in 2024. A schematic of the experiment and beamline at this stage is shown in Figure 4.

4. Future of the Electron Linac as a Multi-user Facility

After 2025, when the ARIEL project is complete, the e-linac must serve its primary purpose of delivering beam to that facility. Several proposed projects and/or e-linac modifications can, however, proceed in parallel without disrupting beam delivery to ARIEL. These include:

DarkLight Phase 2: With the addition of a septum magnet and RF deflector to the beamline setup for DarkLight Phase 1, it will be possible to direct alternate electron bunches to the ARIEL targets and to the experimental site in front of a new high-powered beam dump built to serve DarkLight. In 2025 and the years immediately after, this site will support DarkLight. After its physics program finishes, the same site could be used for other experiments in future. The layout of DarkLight in Phase 2 is shown in Figure 5.

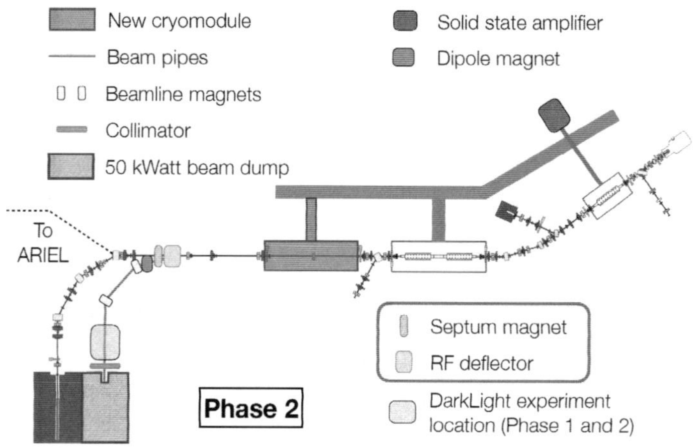

Figure 5. Schematic layout of the DarkLight experiment in phase 2, when it will be able to operate in parallel with ARIEL.

TeraHz: Due to its parameters (CW beam, 30 MeV) the ARIEL e-linac is an ideal driver for a high-intensity THz near-infrared (IR) photon source. There are only a few similar accelerators in the world. The project, led at TRIUMF by Victor Verzilov, is part of the proposed National IR free electron laser program led by the University of Waterloo. Producing THz light requires sub-mm high-charge electron bunches, and some THz production stations (optical transition radiation or synchrotron radiation). A project has been proposed with the objective of demonstrating production of high intensity (MV/cm) broadband radiation, with the ultimate goal of establishing a users' community.

Energy Recovery Linac: Finally, an upgrade of the e-linac to an Energy Recovery Linac (ERL) has been envisaged since the inception of ARIEL. A conceptual design has been developed involving an energy boosting arc [9]. This design can fit alongside the new DarkLight beamline in the e-hall with no anticipated space conflicts and is compatible with the needs of the THz program as well. A schematic of the ERL layout is shown in Figure 6.

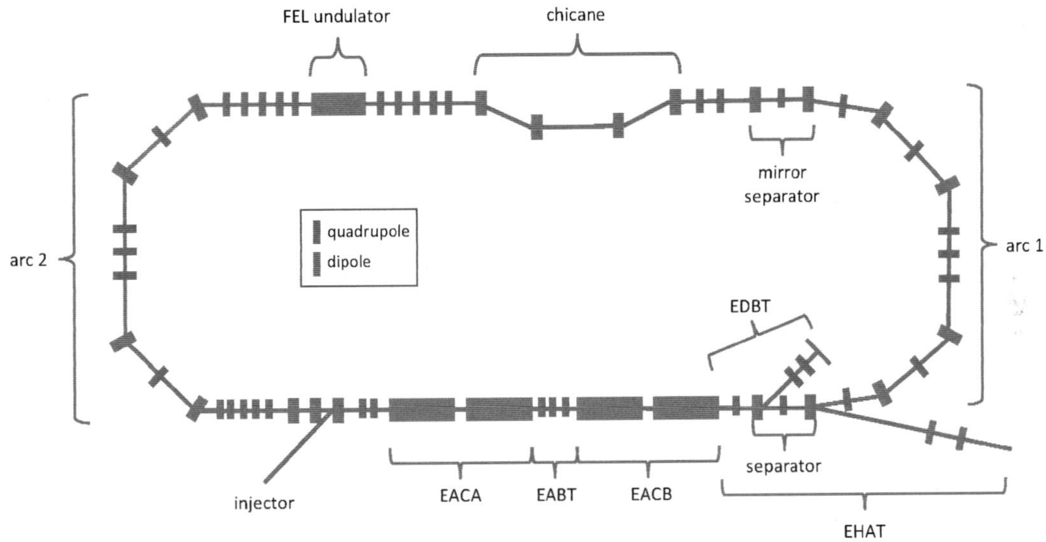

Figure 6. Schematic layout of the ERL upgrade to ARIEL.

References

[1] F. Ames et al., *The TRIUMF ARIEL RF Modulated Thermionic Electron Source*, in *28th International Linear Accelerator Conference*, p. TUPRC020, 2017, DOI.

[2] R. Laxdal et al., *The 30MeV Stage of the ARIEL e-linac*, in *Proc. of International Conference on RF Superconductivity (SRF'17), Lanzhou, China, July 17-21, 2017*, no. 18 in International Conference on RF Superconductivity, (Geneva, Switzerland), pp. 6–12, JACoW, Jan., 2018, DOI.

[3] Y. Ma et al., *First RF Test Results of Two-Cavities Accelerating Cryomodule for ARIEL eLinac at TRIUMF*, in *9th International Particle Accelerator Conference*, 6, 2018, DOI.

[4] Y. Ma, P. Harmer, D. Lang, R. Laxdal, B. Waraich and V. Zvyagintsev, *High Power Coupler Test for ARIEL SC Cavities*, in *Proc. of International Conference on RF Superconductivity (SRF2015), Whistler, BC, Canada, Sept. 13-18, 2015*, no. 17 in International Conference on RF Superconductivity, (Geneva, Switzerland), pp. 1390–1393, JACoW, JACoW, Dec, 2015, DOI.

[5] V. Zvyagintsev et al., *Nine-cell Elliptical Cavity Development at TRIUMF*, in *Proceedings SRF 2011 Conference, Chicago, IL*, 2011.

[6] L. Egoriti, M. Cervantes, T. Day Goodacre and A. Gottberg, *Material development towards a solid 100 kW electron-gamma converter for TRIUMF-ARIEL*, *Nucl. Instrum. Meth. B* **463** (2020) 232 [1907.09363].

[7] N. Esplen, L. Egoriti, B. Paley, T. Planche, C. Hoehr, A. Gottberg et al., *Design optimization of an electron-to-photon conversion target for ultra-high dose rate x-ray (FLASH) experiments at TRIUMF*, *Physics in Medicine & Biology* **67** (2022) 105003.

[8] A.J. Krasznahorkay et al., *Observation of Anomalous Internal Pair Creation in Be8 : A Possible Indication of a Light, Neutral Boson*, *Phys. Rev. Lett.* **116** (2016) 042501 [1504.01527].

[9] C. Gong, *A novel optimization platform and its applications to the TRIUMF energy recovery linac*, Ph.D. thesis, University of British Columbia, 2015. http://dx.doi.org/10.14288/1.0221373.

Low-energy electron scattering facilities in Japan

Toshimi Suda

Research Center for Electron-Photon Science, Tohoku University, 1-2-1 Mikamine, Sendai, 982-0826, Japan

E-mail: suda@lns.tohoku.ac.jp

Abstract. This article reports our research activities of low-energy electron scattering facilities in Japan: the ULQ2 facility for proton charge measurements at Tohoku and the SCRIT facility for the exotic nuclei at RIKEN.

1. Introduction

We currently operate two low-energy electron scattering facilities in Japan for nuclear physics. They are

(i) the ULQ2 facility at Tohoku: $E_e = 10 - 60$ MeV,

(ii) the SCRIT facility at RIBF/RIKEN: $E_e = 150 - 300$ MeV

The ULQ2 facility aims to accurately determine the proton charge radius by measuring the elastic cross section of the proton at very low energies ($E_e = 10 - 60$ MeV), which covers the lowest-ever momentum transfer region.

The SCRIT facility is the first and currently the only electron scattering facility worldwide dedicated for short-lived exotic nuclei.

This report will discuss their details and ongoing physics programs. In addition, a new physics opportunity with such a low-energy electron scattering is discussed.

2. ULQ2 (Ultra-low Q2) facility at Tohoku

2.1. Proton charge radius

Protons are the building blocks of visible matter in the universe. Since their discovery, the proton, and also hydrogen atom as well, have been the most important and central targets for modern physics. For instance, the hydrogen atom was the "laboratory" for establishing quantum mechanics followed by electromagnetic quantum dynamics (QED). The recognition of the finite size of the proton by Hofstadter and his collaborators by electron scattering[1] prompted further studies of its internal structure by using deep inelastic electron scattering, leading to the discovery of quarks and their dynamics, quantum chromodynamics (QCD).

Since Hofstadter's pioneer work, the proton charge radius has continuously been long measured by electron scattering. This is because the radius is one of the fundamental quantities of the proton. In the 1990s, the proton size extraction by hydrogen spectroscopy began, taking advantage of the fact that the orbital energy of the s-wave electron depends on the finite size of the proton. Since their results appeared consistent with the radius determined by electron scattering, the proton radius was widely considered to be ~ 0.88 fm[2].

However, new results from μ-hydrogen spectroscopy in 2010 called this situation into question[3], reporting a radius of 0.844 fm, seriously inconsistent with those determined by electrons thus far. They measured the Lamb shift, which provides the direct access to the proton radius.

This problem called the "proton radius puzzle" [4], majorly impacts nuclear physics because the radius is one of the fundamental quantities of the proton. It also impacts atomic physics since the Rydberg constant and the proton radius are directly connected. Moreover, speculation linking the discrepancy of the radius measured by electron and muon to an unknown difference in their physical properties[5], which should be "beyond the standard model", has received much attention.

Since then, experiments with state-of-the-art technologies have been planned or conducted worldwide, such as electron scattering at JLab and hydrogen spectroscopy experiments in many laboratories[6, 7, 8, 9, 10]. Some new results have been published recently. Some claim to align with those obtained by μ-hydrogen spectroscopy, while others do not. We do not know the reason for the disagreement with the previous results, the situation appears even more confusing.

Here, it is worth mentioning a non-negligible model dependence when extracting the radius from measured electron scattering data has been critically pointed out by I. Sick[11]. In order to realize all requirements discussed in [11], we proposed a ULQ2 electron scattering experiment at the Research Center for Electron-Photon Science (ELPH) of Tohoku University, Japan. After few-year construction efforts, we recently started the measurement.

2.2. ULQ2

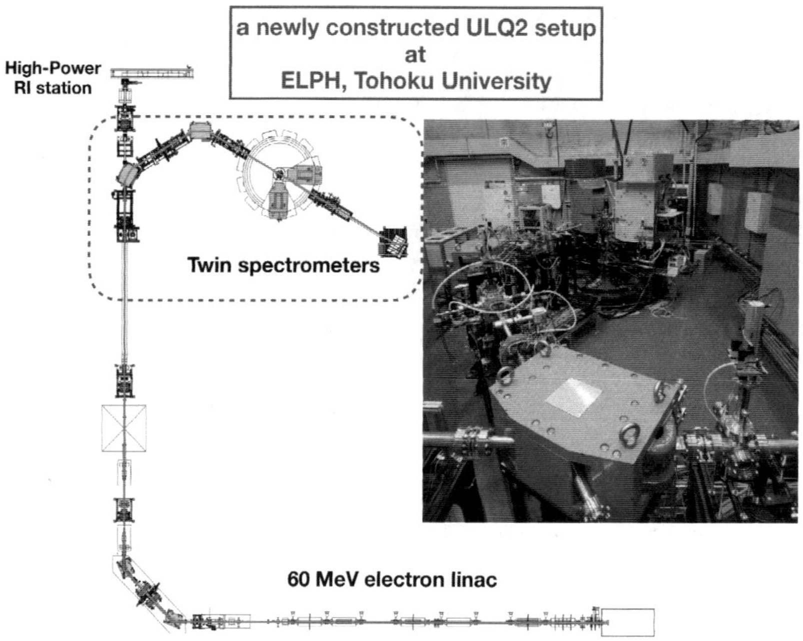

Figure 1. A newly installed ULQ2 setup at ELPH, Tohoku University aiming at determining the proton radius by low-energy electron scattering.

The proton charge radius, $< r_p >$, is defined as[12]

$$\langle r_p^2 \rangle \equiv -6 \, \frac{\mathrm{d}G_E(Q^2)}{\mathrm{d}Q^2}|_{Q^2 \to 0} \ , \qquad (1)$$

here $G_E(Q^2)$ is the proton charge form factor. Q^2 is the four-momentum transfer,

$$Q^2 \equiv \vec{q}^2 - \omega^2 = 4\,E_e\,E_e'\,\sin^2(\theta/2). \tag{2}$$

ω、$\vec{q}$ are the energy and three-momentum transfer; E_e, E_e' are electron energies before and after scattering; and θ is the electron scattering angle.

We designed the ULQ2 setup to realize the measurements and determine the proton radius based on Eq.(1) in the least model-dependent way,

(i) the *absolute* cross section with 10^{-3} accuracy,
(ii) the cross section covering the *lowest-ever* Q^2 region,
(iii) the *Rosenbluth-separated* $G_E(Q^2)$ with 10^{-3} accuracy.

A new beamline and spectrometer system were constructed downstream of a 60-MeV electron linac of ELPH, shown in Fig. 1, to meet these requirements. With an electron beam of $E_e = 10 - 60$ MeV, one covers the scattering angle of $\theta_e = 30 - 150°$, corresponding to $Q^2 = 0.0003 \leq Q^2 \leq 0.008$ (GeV/c^2). We employed CH_2 as a target to determine the **absolute** electron-proton elastic cross section with an accuracy of 10^{-3}. Since the charge radius of ^{12}C has an accuracy of 10^{-3}, it is possible to determine the cross section of the proton with high precision by relative elastic scattering measurements of ^{12}C.

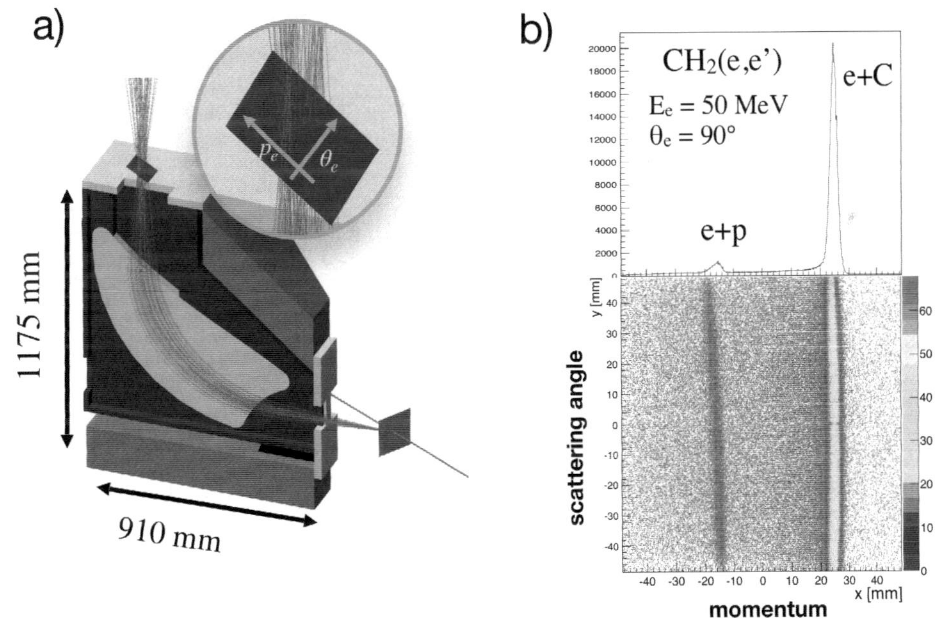

Figure 2. a) A newly constructed electron spectrometer. b) Electron momentum spectrum and 2D plot of CH_2(e,e').

The construction of the ULQ2 facility, consisting of a new beamline and high-resolution twin spectrometers, was recently completed, followed by their commissioning. One spectrometer is used for measuring the angular distribution of elastic scattering to extract $G_E(Q^2)$ using the Rosenbluth technique, and the other is used to monitor the C/H ratio under a fixed Q^2. Note that the C/H ratio is known to change with beam irradiation.

The spectrometer is a vertical-bending magnetic type, as shown in Fig. 2 a), which allows one to measure the momentum and scattering angle of scattered electrons, and cover a wide range of scattering angles of $\theta = 30 - 150°$. The spectrometer was carefully designed so that the two-dimensional (2D) position measurements using a detector at the focal plane would provide the momentum and scattering angle of the scattered electrons with the required accuracy. The focal plane detectors are two single-sided silicon detectors (SSSD) developed in collaboration with the JPARC muon g-2 group[13].

Figure 2 b) shows a 2D distribution at the focal plane of the scattered electrons from CH_2 measured with $E_e = 50$ MeV and $\theta_e = 90°$. Thanks to the low-energy electron scattering, elastically scattered electrons from carbon and proton were both in the spectrometer's momentum bite of 11 %, which allowed the determination of the absolute electron-proton scattering cross section relative to that of carbon.

The transfer matrix elements of both spectrometers were fully determined up to the second order and confirmed to agree with the designed values. We have already started a series of physics runs in July 2022.

3. The SCRIT facility at RIBF/RIKEN

3.1. Electron scattering for exotic nuclei

Following the pioneering work of Hofstadter and his collaborators in revealing the size and shape of the subatomic system[14], electron scattering has been widely recognized as one of the most powerful tools for studying the internal structure of atomic nuclei. Indeed, it has consistently played an essential role in the structure study of nuclei. We must highlight, however, that only stable nuclei have been so far studied, and short-lived nuclei have never been investigated by electron scattering.

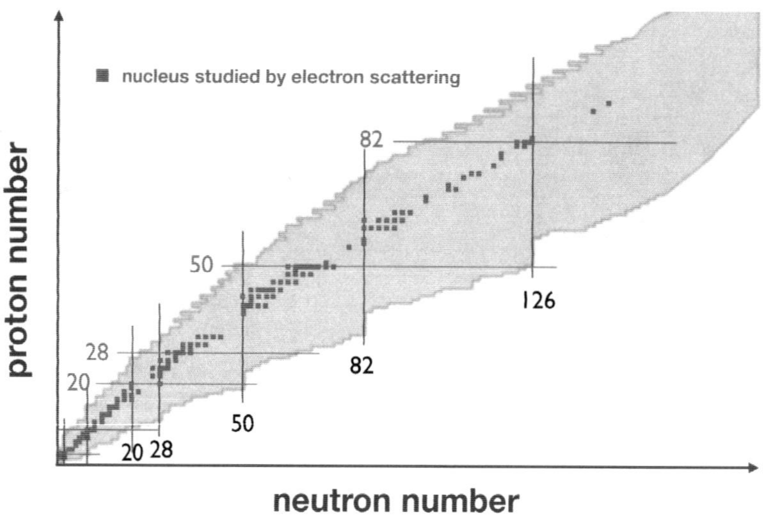

Figure 3. Nuclei studied by elastic electron scattering[15] are indicated by red squares. Magic numbers known for stable nuclei are indicated by black (red) lines for neutrons (protons).

Figure 3 shows a nuclear chart indicating nuclei studied by electron scattering[15]. One may immediately notice that the nuclei targeted for electron scattering were stuck to the "valley of stability." Strictly speaking, however, there are a few examples of unstable nuclei, such as ^{3}H,

^{14}C, ^{41}Ca, where their lifetimes are, without exception, at least more than years long enough to prepare a "thick" target for electron scattering.

One of the hot issues of modern nuclear physics is to explore exotic nuclear structures found or suggested in nuclei far from stability. This is the reason why high-power heavy-ion accelerators and/or high-power driver accelerators were constructed or are under construction worldwide, which allow accessing nuclei further from stability.

3.2. SCRIT electron scattering facility

When we began constructing the RIKEN RI (Radioactive-Isotope) Beam Factory in the 2000s to study the structure and reaction of exotic short-lived nuclei, we started dreaming of an electron scattering facility dedicated for short-lived exotic nuclei.

Obviously, a key parameter to realize electron scattering for short-lived nuclei is the collision luminosity, L, defined as follows:

$$\frac{\mathrm{d}N}{\mathrm{d}t} = L \cdot \sigma, \tag{3}$$

where $\mathrm{d}N/\mathrm{d}t$ is the scattering rate per unit time and σ is the cross section. Detailed simulations resulted in the required luminosity for elastic electron scattering to be at least $L \sim 10^{27}(/\mathrm{cm}^2/\mathrm{s})$. With this luminosity, measuring the charge form factor of medium-heavy exotic nuclei covering the first diffraction maximum is possible, which provides the size and shape of the charge density distributions of exotic nuclei[16].

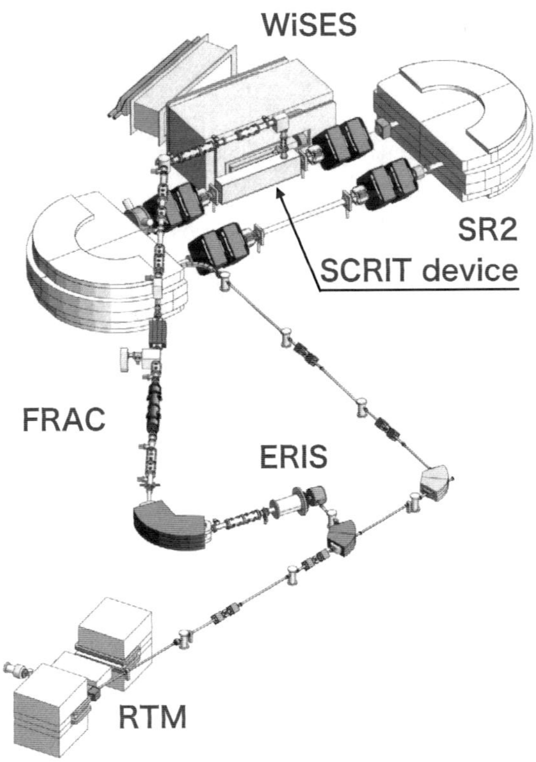

Figure 4. SCRIT electron scattering facility in the RIKEN RI Beam Factory.

We have constructed the world's first electron scattering facility at the RIKEN RI Beam Factory in Japan that employs a novel target forming technique[18]; we invented and named SCRIT (Self-Confined RI Ion Target), to achieve the required luminosity for elastic scattering, $L \sim 10^{27}$ /cm^2/s with a small number of target nuclei, such as $N \sim 10^{7-8}$.

Figure 4 shows the floor layout of the SCRIT electron scattering facility, which comprises an injector (RTM: Race-Track Microtron), a 700-MeV electron storage ring (SR2: SCRIT-equipped RIKEN storage ring), a SCRIT device, and an electron spectrometer (WiSES: Window-frame Spectrometer for Electron Scattering). Neutron-rich unstable nuclei are produced via photo-fission of ^{238}U at an ISOL (ERIS: Electron-beam-driven RI separator for SCRIT), followed by a cooler-buncher (FRAC: Rringing-Rf-field-Activated dc-to-pulse Converter). The pulsed ions are then transported and injected into the circulating electron beam in SR2. The scattered electrons from the trapped unstable nucleus are detected by WiSES, consisting of a window-frame dipole magnet sandwiched by a pair of drift chambers. Note that the trapping period of the externally injected ions of unstable nuclei is easily adjusted by controlling the SCRIT electrodes in order to keep the purity of the trapped target.

The performance study of the SCRIT facility using $N \sim 10^8$/s of a stable ^{132}Xe[19], where the trapping period was a few seconds, successfully demonstrated that electron scattering for unstable nuclei is feasible. We are currently conducting a series of elastic scattering measurements for xenon (Xe) isotopes from ^{124}Xe ^{138}Xe, and $N = 82$ isotones, which include unstable nuclei of ^{137}Cs and ^{138}Xe. These measurements are the first electron scattering experiments targeting online-produced unstable nuclei.

4. A new physics case: neutron-distribution radius of atomic nuclei

It is a textbook example that electron elastic scattering reveals how the charge density spatially distributes in stable nuclei. Since the proton charge dominates the charge density of a nucleus, the proton distributions of various nuclei have been precisely determined so far. It is well known that they were keys for establishing the modern nuclear-structure models.

Since the neutron charge is zero, less knowledge of the neutron distribution has been obtained. Exceptional examples are parity-violating electron scattering of ^{208}Pb using polarized electrons recently conducted at JLab[20, 21] aimed at measuring the radius of the neutron distribution, and traditional magnetic electron scattering to probe a valence neutron distribution[22]. Unfortunately, both experiments require extremely high luminosity due to small magnetic-scattering cross sections and weak-interaction effects, making it hard to imagine their application to exotic nuclei, where extremely low luminosities should be expected.

Recently, a new way to access neutron distributions by electron scattering has been proposed[23], which may overcome such difficulties mentioned above. Kurasawa and Suzuki have shown by their full relativistic nuclear-structure models that the fourth moment of the nuclear charge density distribution includes the radius of the neutron distribution explicitly.

Here, the nth-moment of the charge density distribution, $\rho_c(r)$, is defined as,

$$< r_c^n > = \int r^n \rho_c(r) \, \mathrm{d}^3 r. \tag{4}$$

For the second moment, $< r_c^2 >$, is expressed as

$$< r_c^2 > = < R_{p(point)}^2 > + r_p^2 + \frac{N}{Z} r_n^2 + \text{rel. corr.}, \tag{5}$$

where $< R_{p(point)} >$, r_p,r_n are the radius of the point proton distribution, the proton charge radius, and the neutron charge radius, respectively. The relativistic correction includes contributions from the spin-orbit interaction. Here, $< r_c^2 >$ is measured experimentally by electron scattering, μ-X ray, and isotope shifts, and $R_{p(point)}$ is calculated theoretically. Since the second moment corresponds to the exact (charge) size of atomic nuclei, providing a strict testing ground for our understanding of the nuclear structure. This is the reason why they have been measured extensively, including those of the unstable nuclei[24].

The fourth moment, $< r_c^4 >$, is shown to include the neutron distribution radius[23] as,

$$< r_c^4 > = < R_{p(point)}^4 > + \frac{10}{3} < R_{p(point)}^2 > < r_p^2 > + \frac{10}{3} < R_{n(point)}^2 > < r_n^2 > \frac{N}{Z} + \text{rel.corr..} \quad (6)$$

One finds that the neutron distribution radius, $< R_{n(point)} >$, appears explicitly in the expression. Hence, assuming $< r_c^4 >$ is measured experimentally and using known values of $< R_{p(point)}^2 >, < r_p^2 >$ and $< r_n^2 >$, it becomes possible to determine $< R_{n(point)} >$ from $< r_c^4 >$ by elastic electron scattering.

We tried applying this theory to a doubly-magic stable nucleus ^{208}Pb using the charge density distribution, $\rho_c(r)$, known precisely by elastic electron scattering. As a result, the neutron distribution radius, $< R_{n(point)} >$, is determined as 5.736 ±0.013fm[25]. Extracting the (point) proton distribution radius, 5.455±0.012 fm, from the experimentally determined charge radius, 5.503±0.002 fm, the neutron skin thickness, defined as the difference between the proton and neutron distribution radii, could then be determined as 0.282±0.024 fm. This result looks surprisingly consistent with the skin thickness recently determined by the parity-violating electron scattering experiment at JLab, 0.283±0.071 fm[20, 21].

The neutron skin thickness of ^{48}Ca is also estimated in the same way, and obtained as 0.219±0.013 fm[25]. The skin thickness of ^{48}Ca determined by the parity-violating electron scattering experiment at JLab was recently reported[26] to be 0.121±0.026 fm.

As previously discussed, electron scattering for neutron-rich exotic nuclei will soon be started at the SCRIT electron scattering facility. However, one may ask whether it is possible to access also the neutron distribution radius of neutron-rich unstable nuclei through $< r_c^4 >$ at the SCRIT facility.

There are two ways to determine $< r_c^4 >$ of the charge density distributions.

(i) First is using equation (4) with $\rho_c(r)$ determined by elastic electron scattering covering several q,

(ii) Second is through the Taylor expansion of the charge form factor, $F_c(q)$, at the low q region,

$$F_c(q) = 1 - \frac{< r_c^2 >}{3!} q^2 + \frac{< r_c^4 >}{5!} q^4 + ... \quad (7)$$

Considering the fact that the elastic cross section has $1/q^4$ dependence, and $< R_{p(point)}^2 >$ of the exotic nuclei are known by isotope shift measurements, method (2) may allow one to study the neutron distribution radius of exotic neutron-rich nuclei even at the low-luminosity electron scattering facility. Based on this finding, a feasibility study on this exciting research opportunity at the SCRIT electron scattering facility is underway.

5. Summary

We have reported the current status and physics program at Japan's low-energy electron scattering facilities: ULQ2 at Tohoku and SCRIT at RIKEN/RIBF, where both facilities are in operation. The physics runs to determine the least model-dependent proton charge radius starts soon at ULQ2, and the world's first electron scattering of exotic nuclei at SCRIT is ongoing.

A new opportunity in nuclear physics at such low-energy electron scattering facilities has also been discussed. It has been pointed out that the radius of the neutron distribution in atomic nuclei may be accessible by the fourth moment of the charge density distribution determined by electron scattering.

Both ULQ2 and SCRIT facilities are unique also in the sense that one measures the charge form factor of stable and exotic nuclei at very low momentum transfer, where one may extract the fourth moment of the charge density distribution.

References

[1] Hofstadter R: Nobel Lecture Archive, https://www.nobelprize.org/prizes/physics/1961/hofstadter/lecture/ (accessed 2022-01-10)
[2] Carlson C 2015 *Prog. Part. Nucl. Phys.* **82** 59-77.
[3] Pohl R *et al.* 2010 *Nature* **466** 213.
[4] Antognini A *et al.* 2013 *Science* **339** 417.
[5] Carlson C.E. and Rislow B. C. 2012 *Phys. Rev.* **D86** 035013.
[6] Xiong W *et al.* 2019 *Nature* **575** 147-150.
[7] Beyer A *et al.* 2017 *Science* **358** 79-85.
[8] Fleurbaey H *et al.* 2018 *Phys. Rev. Lett.* **120** 183001.
[9] Bezginov N *et al.* 2019 *Science* **365** 1007-1012.
[10] Grinin A*et al.* 2020 *Science* **370** 1061-1066.
[11] Sick I *Atoms* 2018 **6** 2.
[12] Miller G. A. 2019 *Phys. Rev.* **C99** 035202.
[13] Yamanaka T *et al.* 2020*Nucl. Instrum. and Methods* A958 162786 .
[14] Hofstadter R 1957 *Ann, Rev. Nucl. Sci.* 7 231.
[15] de Vries H, de Jager C.W and de Vries C 1987 *At. Data Nucl. Data Table* 36 495.
[16] Suda T and Simon H 2017 *Prog. Part. Nucl. Phys.* 96 1-31.
[17] Suda T 2020 *Journal of Physics*, 1643 012159.
[18] Wakasugi M, Suda T and Yano Y 2004 *Nucl. Instrum. Methods* A 532 216.
[19] Tsukada K *et al.* 2017 *Phys. Rev. Lett.* 118 262501.
[20] Abrahamyan S *et al.* 2012 *Phys. Rev. Lett.* 108 112502.
[21] Adhikari D *et al.* 2021 *Phys. Rev. Lett.* 126 172502.
[22] Donnelly T W and Sick I 1984 *Reviews of Modern Physics*, 56 461.
[23] Kurasawa H and Suzuki T 2019 *Prog. Theor. Exp. Phys.* 2019 113D01.
[24] Campbell P, Morre I.D. and Pearson M.R 2016 *Prog. Part. Nucl. Phys.* **86** 127-180.
[25] Kurasawa H, Suda T and Suzuki T 2021 *Prog. Theor. Exp. Phys.* 013D02.
[26] Adhikari D *et al.* *arXiv:2205.11593*

Acknowledgments

The author would like to thank all the members of the ULQ2 and SCRIT collaborations. This work was supported by Grants-in-Aid for Scientific Research (S)(Grands No. 22224004, JP16H06340, JP20H05635).

TRIUMF-2022 IOP Publishing
Journal of Physics: Conference Series **2391** (2022) 012005 doi:10.1088/1742-6596/2391/1/012005

The MESA science program: dark matter and more

Luca Doria
PRISMA^{+} Cluster of Excellence and Johannes Gutenberg Univeristy Mainz
Institute for Nuclear Physics
J.-J.-Becher-Weg 45, 55128, Mainz, Germany

E-mail: doria@uni-mainz.de

Abstract. The Mainz Energy-recovery Superconducting Accelerator MESA will allow precise measurements in hadron and nuclear physics, as well as exciting opportunities in dark matter searches. Three experiments will be built around this new and unique facility: MAGIX, P2, and DarkMESA. In this contribution, the MESA science program will be presented, with focus on dark matter and new physics searches.

1. Introduction

In this contribution, the physics program of MESA will be discussed. **MESA** is a new experimental facility centered around the **M**ainz **E**nergy-recovery **S**uperconducting **A**ccelerator. The experiments at MESA will cover a wide program from dark matter searches to nuclear astrophysics.

A variety of astrophysical and cosmological observations point to the existence of Dark Matter (DM). In the last decades, an effort was made for searching for DM with particle colliders, fixed-target experiments, indirect detection techniques, and underground direct detection experiments. Still, the nature of DM remains elusive, and more sensitive and diversified experiments are required. A large class of models describe DM as a relic from the early Universe where DM was in thermodynamic equilibrium with Standard Model (SM) particles. In these models, the DM abundance is set when its annihilation rate in SM particles became smaller than the expansion rate of the universe in a process known as *freeze-out*. While this mechanism is very compelling, it allows for a broad range of DM masses (keV/c^2-TeV/c^2) and interaction cross sections. The prime target of many experiments is a DM candidate called WIMP (Weakly-Interacting Massive Particle), motivated by models of physics beyond the SM (most notably supersymmetric models). This range can be effectively tackled by high-energy colliders and direct detection experiments. In the light dark matter (<GeV/c^2) range (LDM), these methods become less effective, and different techniques are needed. LDM production at accelerators can in principle overcome the limitations of direct detection experiments (the detection energy threshold) by producing DM particles with enough momentum to be detected, even if they are relatively light. High-energy colliders are not optimised for the low-mass range, and lower energy accelerators coupled with fixed targets can have an advantage in LDM searches. In this context, low-energy accelerator searches can also contribute to the "X17 puzzle" [1] or in the search of axion-like particles (ALPs). On top of direct new physics searches, a high intensity, low-energy electron accelerator has the strong potential to contribute to precision measurements in hadron and nuclear physics. A prime example is the opportunity to measure the nucleon form factors

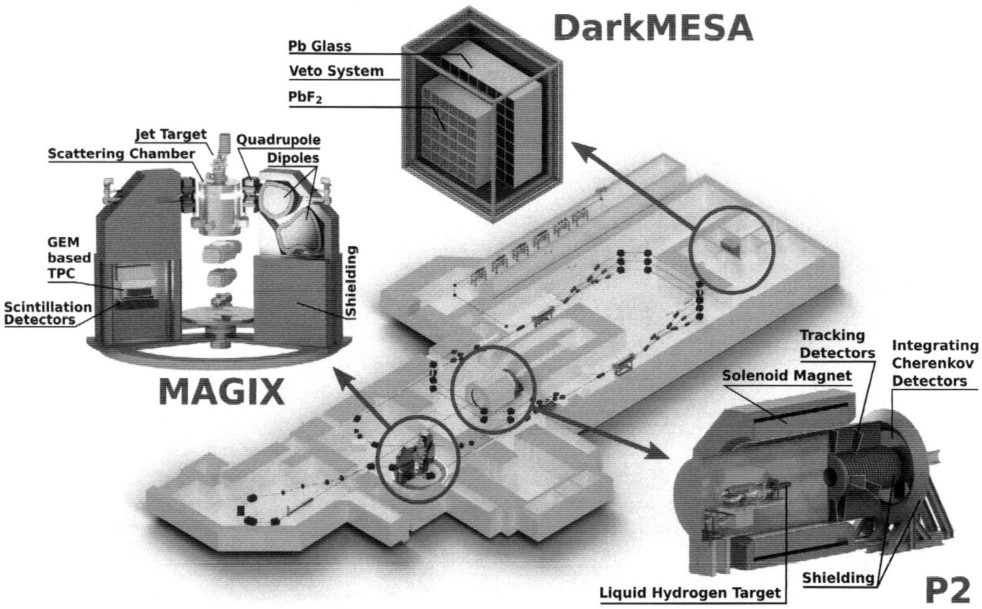

Figure 1. The MESA accelerator floor plan with the three experiments: DarkMESA, MAGIX, and P2.

at low momentum transfer, clarifying recent disagreements between electron scattering and spectroscopy measurements. Another research direction consists in the precision measurements of nuclear cross sections of few-body nuclear systems for elucidating properties of the nuclear potential, also in connection with effective field theories. Furthermore, MESA can contribute to the measurement of S-factors for nuclear reactions of astrophysical interest.

2. The MESA accelerator

The Institute for Nuclear Physics at the Mainz University is building a new multi-turn, 100% duty-cycle, energy recovery linac for precision experiments with a beam energy of 100-200 MeV [2] (see Fig. 1). MESA (Mainz Energy-Recovering Superconducting Accelerator) will operate in two modes: energy recovery mode (ERM) and external beam mode (XBM).
In ERM, the accelerator will provide a beam current of up to 1 mA at 105 MeV for the MAGIX internal target experiment with multi-turn energy recovery capability. In XBM, a polarised beam of 150 μA will be provided to the P2 experiment [3]. An additional experiment, DarkMESA, will be placed after the P2 beam dump, running in parallel to P2 and searching for LDM particles.
The MESA accelerator will consist of a polarised source followed by a low energy beam transport system containing a spin manipulation system and a chopper-buncher section. Normal conducting cavities will accelerate the beam up to 5 MeV before injection into the main linac equipped with a total of four ELBE-like 9-cell superconducting cavities [4] installed in two cryomodules. The linac will provide an energy gain of 50 MeV/pass.

3. Light Dark Matter

In this section, we briefly introduce a simple model for light dark matter, which also constitute a benchmark for the sensitivity of experiments in this field [5, 6].

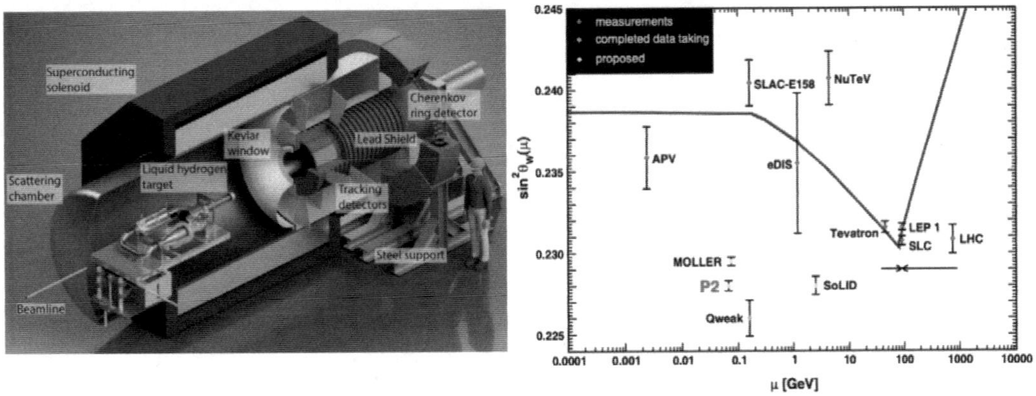

Figure 2. Left: Technical view of the P2 experiment. **Right:** Evolution with the energy scale of the Weinberg angle together with the existing experimental results and the planned experiments.

In the <GeV/c^2 range, for retaining a thermal origin of DM, the existence of additional interactions has to be postulated. For small masses, considering only electroweak-scale cross sections, the annihilation rate would not be sufficient, leading to DM overproduction. A relevant class of LDM models is based on the idea that DM particles belong to a *dark sector* interacting with the SM via one (or more) mediator particle(s). Dark sector models can be classified by the type of mediator particle (the *portal*) and the type of DM particle. In general, the dark sector might contain more mediators and particles. A simple model which captures the essence of dark sector physics contains a massive vector mediator particle (a "dark photon") and one DM particle. The model's Lagrangian is

$$\begin{aligned}\mathcal{L} \supseteq & -\frac{1}{4}F'_{\mu\nu}F'^{\mu\nu} + \frac{\epsilon_Y}{2}F'_{\mu\nu}B^{\mu\nu} \\ & + \frac{m^2_{A'}}{2}A'_\mu A'^\mu + g_D A'_\mu J^\mu_\chi + g_Y B_\mu J^\mu_Y \quad ,\end{aligned} \tag{1}$$

where $F'_{\mu\nu} = \partial_\mu A'_\nu - \partial_\nu A'_\mu$ and $B_{\mu\nu} = \partial_\mu B_\nu - \partial_\nu B_\mu$ are the dark photon and the hypercharge fields, g_D is the dark gauge coupling, and J^μ_χ and J^μ_Y the DM and hypercharge currents, respectively. After electroweak symmetry breaking, the dark photon mixes with the SM photon and the Z boson

$$\frac{\epsilon_Y}{2}F'_{\mu\nu}B^{\mu\nu} \to \frac{\epsilon}{2}F'_{\mu\nu}F^{\mu\nu} + \frac{\epsilon_Z}{2}F'_{\mu\nu}Z^{\mu\nu} \quad , \tag{2}$$

where $\epsilon = \epsilon_Y/\cos\theta_W$, $\epsilon_Z = \epsilon_Y/\sin\theta_W$, and θ_W is the weak mixing angle. After diagonalization, the coupling of the dark photon with DM and the SM photon is

$$g_D A'_\mu J^\mu_\chi + g_Y B_\mu J^\mu_Y \to A'_\mu(g_D J^\mu_\chi + \epsilon e J^\mu_{EM}) \quad , \tag{3}$$

where J^μ_{EM} is the SM electromagnetic current. The form of the DM current J^μ_χ depends on the exact nature of the DM particle. The coupling of the dark photon to SM particles happens via the coupling constant ϵe, while the dark fine structure constant $\alpha_D = \sqrt{4\pi g_D}$ describes the coupling with DM.
It is useful to define the dimensionless combination of the model parameters

$$y = \epsilon^2\alpha_D\left(\frac{m_\chi}{m_{A'}}\right)^4 \quad , \tag{4}$$

which is proportional to the thermally averaged DM annihilation rate.

4. The P2 Experiment

The main aim of the P2 experiment [3] is the measurement of the parity violating asymmetry

$$A^{PV} = \frac{d\sigma_{ep}^{+} - d\sigma_{ep}^{-}}{d\sigma_{ep}^{+} + d\sigma_{ep}^{-}} \quad , \tag{5}$$

where $\sigma_{ep}^{\pm}$ is the elastic electron-proton cross section with electrons of helicity $\pm 1/2$ respectively. The observable A^{PV} depends on the weak form factor of the proton $Q_W(p) = 1 - 4\sin^2\theta_W$ (neglecting radiative corrections) and thus allows the extraction of the Weinberg (or weak) angle θ_W. The Weinberg angle is a key parameter of the Standard Model and its dependence on the energy scale can be precisely calculated (see Fig. 2, right). P2 aims at a $\sim 0.15\%$ precision measurement of $\sin^2\theta_W$ complementing the other precise measurement at colliders made at the Z-pole. Measuring a deviation from the Standard Model predictions would be a signal for the presence of new physics.

P2 employs a cryogenic target placed at the center of a superconducting solenoidal magnetic field. Elastically scattered electrons are measured by 82 fused-silica bars. Given the high-precision goal of the experiment and the need for keeping a reasonable measuring time, electrons are not counted as single events, but the silica bars will measure an asymmetry in the output current (the rate will be of the order of 10^{12} Hz in the detector). Tracking is provided by High-Voltage Monolithic Active Pixel Sensors (HV-MAPS, [7]) which, in combination with the magnetic field, will measure the momentum of the tracks. A technical drawing of the P2 detector is shown in Fig. 2 (left).

5. The MAGIX Experiment

The **MA**inz **G**as **I**njection Target **EX**periment **(MAGIX)** is a flexible experiment exploiting the unique combination of a supersonic gas-jet target and the MESA continuous beam in energy recovery mode. The experiment is based on two magnetic spectrometers ("STAR" and "PORT") which can be operated in coincidence with momentum resolution $\delta p/p \sim 10^{-4}$ and low-material budget focal plane tracking detectors (Fig. 1). The tracking detectors are time projection chambers (TPCs) with GEM (Gas Electron Multiplier) readout. To shape the drift field and keep the material budget as low as possible (at low momenta, multiple scattering degrades resolution), field shaping extensions are used instead of a more common field cage on the entrance face of the TPC. A new approach is adopted for online calibration by employing a 266 nm laser system to release ionisation electrons from fixed positions on the cathode for producing a snapshot of the dynamic field distortions.

MAGIX will feature a gas-jet target [8] which will allow beam recirculation in MESA. The jet target was already successfully tested at the A1 collaboration setup with hydrogen and argon as targets [9].

MAGIX will allow precision measurements in a variety of fields ranging from hadron physics to nuclear astrophysics and dark sector searches.

Dark Matter.

For LDM searches, candidate particles are for example dark photons introduced in Sec. 3. The dark photon can be produced through a mechanism similar to bremsstrahlung on a heavy nuclear target Z via the reaction $e^-Z \to e^-Z\gamma'$ or e^+e^- annihilation. If the dark photon decays into SM particles, *e.g.* $\gamma' \to e^+e^-$, the electron/positron final state can be detected in coincidence in the two spectrometers and a peak-search on the QED background can be performed [10, 11]. A similar strategy can be employed for the search for the X17 particle or axions.

If the dark photon decays invisibly (*e.g.* into light dark matter particles $\gamma' \to \chi\bar{\chi}$), this will require the measurement of the recoil target nucleus for fully reconstructing the kinematics. A peak-search on the reconstructed missing mass $m_{\gamma'}^2 = (p_{beam} - p_Z - p_{e'})^2$ will be performed in

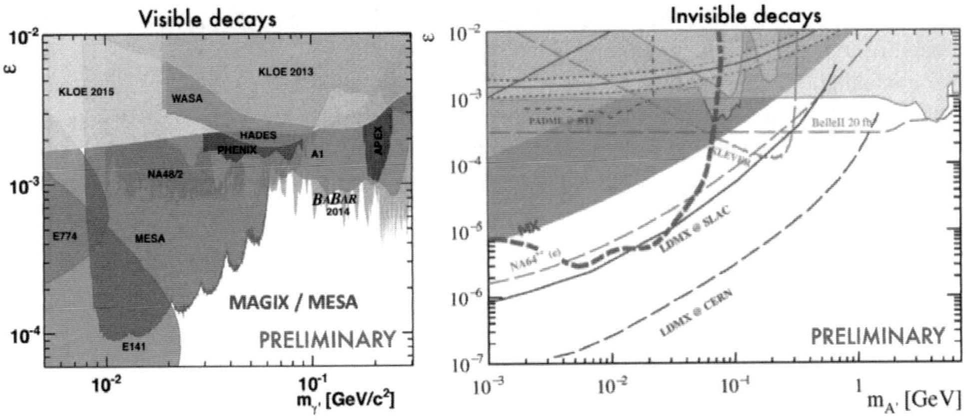

Figure 3. Simulated exclusion limits for the MAGIX experiment where a dark photon γ' is produced by the bremsstrahlung process on a heavy nucleus Z: $eZ \to eZ\gamma'$.
Left: Exclusion limits for the dark photon visible decay $\gamma' \to e^+e^-$. **Right:** Exclusion limits for the dark photon invisible decay $\gamma' \to \bar{\chi}\chi$ where χ is a fermionic light dark matter particle.

this case with the addition of a silicon detector for detecting the recoil proton or nucleus.
A preliminary simulation of the exclusion limits for dark photon decays achievable with MAGIX is shown in Fig. 3.

Form Factors.

Form factors are a fundamental quantity describing the low energy properties of the nucleon. They can be accessed through electron scattering measuring the cross section

$$\frac{d\sigma}{d\Omega} = \left(\frac{d\sigma}{d\Omega}\right)_{\text{Mott}} \frac{G_E^2(Q^2) + \tau G_M^2(Q^2)}{\epsilon(1+\tau)} \quad , \tag{6}$$

where ϵ is the virtual photon polarisation, $\tau = Q^2/(4m_p^2)$, with m_p the mass of the proton, $Q^2 = -q^2$ the squared momentum transfer, and G_E, G_M the Sachs form factors (the electric and magnetic form factors, respectively). $(d\sigma/d\Omega)_{\text{Mott}}$ is the Mott cross section.
In particular, the radius of the proton can be derived from

$$\langle r_p \rangle = -\frac{6}{G(0)} \frac{dG_E(Q^2)}{dQ^2}|_{Q^2=0} \quad . \tag{7}$$

In recent years, the determination of the radius from electron scattering [12] became inconsistent with the high-precision number extracted from spectroscopic measurements of muonic atoms [13]. The disagreement became known as the "proton radius puzzle".
MAGIX can perform a high-precision measurement of the cross section in Eq. 6 focusing on low-momentum transfers in a region very sensitive to the value of r_p. The new measurements have the potential for elucidating the disagreement of electron scattering analyses with the spectroscopy value.

Nuclear Astrophysics.

Stellar burning is driven by nuclear fusion reactions in the core of the stars. Nuclear fusion starts from the lightest elements (hydrogen, helium), and builds heavier nuclei. The nucleosynthesis process $^{12}C(\alpha,\gamma)^{16}O$ is of key importance since its S-factor is a major parameter for determining the abundance ratio of oxygen to carbon in stellar nucleosynthesis and moreover, it plays a crucial role in the pre-supernova star evolution. At MAGIX, the idea is to measure the

electro-disintegration reaction $^{16}O(e,e'\alpha)^{12}C$, scattering an electron on oxygen and detecting in coincidence the α particle. This inverse kinematics reaction is connected with the stellar one only by kinematical factors dependent on the angular momentum of the involved particles [14]. The electro-disintegration has some advantages over the exact inverse photo-disintegration reaction $^{16}O(\gamma,\alpha)^{12}C$. An advantage is that the background from cosmic rays is strongly suppressed by the coincidence measurement of the electron and the recoil α. Furthermore, the virtual photon flux is higher and can be fixed by choosing electron scattering angle and beam energy. Another advantage is that the energy resolution of the (virtual) photon can be precisely measured detecting the scattered electron in a high-resolution detector.
A relevant disadvantage is that the longitudinal and interference terms of the electron-scattering reaction are currently unknown. These terms must be determined experimentally and employed in the extrapolation to small energies. Another disadvantage is that target gas contamination by other isotopes (^{17}O, ^{18}O, ^{14}N) causes additional backgrounds.
The simulations predict that MAGIX will be able to measure the S-factor for center of mass energies above 1 MeV with a precision similar to the existing data [15]. With the addition of a zero-degree tagger (ZDT), it would be possible to detect the scattered electrons with a scattering angle $< 0.5°$. The ZDT detector will exploit a MESA dipole downstream of the MAGIX beamline and will also play the role of luminosity monitor for the whole experimental program.
Few-Body Systems.
Modern *ab initio* nuclear physics aims at a coherent description of few- and intermediate-body nuclei solving the quantum nuclear problem exactly together with the use of potentials derived with effective field theory (EFT).
EFT allows for theoretically controlled calculations where the strong force among nuclei contains interactions beyond the nucleon-nucleon coupling, such as the three-nucleon force. These techniques can be extended to intermediate-mass nuclei, also through the use of cluster EFT, where nuclei can be viewed as bound systems of α particles.
Given the perturbative nature of the electromagnetic force, electron accelerators are very well suited for the investigation of nuclear systems since the electron interaction part is completely known and disentangled from the strong force component to be studied.
MAGIX will allow the precise study of nuclear systems like ^2H, $^{3,4}He$ and the "cluster nucleus" ^{12}C for validating EFT theories and modern *ab initio* codes for the solution of the quantum many-body problem.

MAGIX is currently in its advanced design phase. The design of the magnets is complete, as well as the rotation system. The conceptual design of a GEM-based low-material budget TPC has been finalised and prototypes are tested in the laboratory. The experimental hall is completed and installation of the first components will start at the end of 2022. Significant progress has been achieved in the simulation and analysis software development, with a focus on the simulation of the dark photon production and astrophysical reactions. A key part of the experiment, the jet target, has been already built and tested at the A1 Collaboration experimental setup [8, 9].

6. The DarkMESA Experiment
DarkMESA is a beam dump experiment currently actively developed [16, 17, 18]. DarkMESA will take advantage of the P2 beam dump installing a calorimeter ~ 23 m downstream of it in a radiation-shielded area. Referring to the simple LDM model described in Sec. 3, a dark photon could be produced in the beam dump (through bremsstrahlung or e^+e^- annihilation) and then decay into pairs of DM particles. DM will be then detected (via dark photon exchange) in the calorimeter.
The advantage of such an experiment is the possibility to investigate both the dark photon

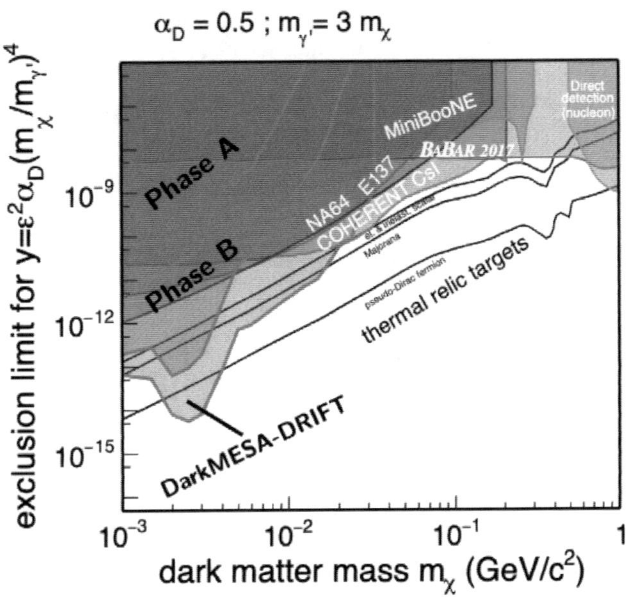

Figure 4. Simulated exclusion limits for the DarkMESA experiment where a dark photon γ' is produced by the bremsstrahlung process and e^+e^- annihilation in the P2 beam dump followed by the decay $\gamma' \to \bar{\chi}\chi$ where χ is a fermionic light dark matter particle. The exclusion limits of the two experimental phases (A,B) are indicated together with the projected limits employing the DRIFT TPC detector.

production and the DM interaction (rescattering in the detector).
The current plan for the construction of the experiment is divided into two phases. Phase-A consists in the construction of a prototype calorimeter module comprised by a matrix of 5×5 PbF_2 Cherenkov crystals for a total volume of $\sim$4 dm^3 surrounded by a hermetic cosmic ray veto system built with two plastic scintillator layers and one lead layer (Fig. 5). The design follows successful tests by the BDX Collaboration, which is proposing a similar experiment at Jefferson Lab [19].
In particular, a dedicated electronics has been developed for the readout of silicon photomultipliers optically coupled to the veto plastic scintillator planes. In Phase-B, the full volume available in the experimental hall will be exploited with the construction of a PbF_2, 0.12 m^3 calorimeter and a lead-glass, 0.58 m^3 calorimeter. The advantage of Cherenkov crystals is their speed and relatively low sensitivity to background neutrons. Exploiting the 2.2×10^{22} (6600 hours) electrons on target delivered to the P2 experiment at 150 μA of beam current, a total charge of $\sim$5400 C will be deposited in the beam dump and will be available for DarkMESA. Besides the calorimeters, it is under study also the opportunity to add a negative-ion TPC to the experimental setup. In particular, the chamber will leverage the technology of the DRIFT experiment optimised for dark matter searches with 20 keV nuclear recoil detection threshold [20].
In Fig. 4, simulated 90% exclusion limits for the two experimental phases and for the DRIFT detector technology are shown.

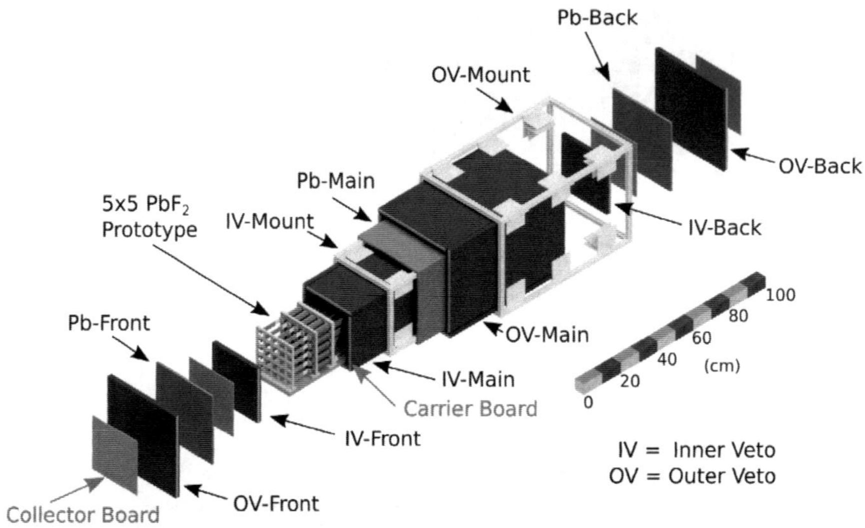

Figure 5. Exploded view of the DarkMESA prototype detector. A matrix of 5x5 PbF_2 Cherenkov crystals is placed inside two hermetic layers of active veto plastic scintillators and one layer of lead. The active shielding will be used for vetoing cosmic rays, while the crystals will be used for DM detection.

6.1. Acknowledgments

The author is indebted to the MAGIX/DarkMESA collaboration and the PRISMA$^+$ Cluster of Excellence at the Johannes Gutenberg University Mainz. He wishes also to thank the organisers of the "New Scientific Opportunities at the TRIUMF ARIEL e-linac" workshop for the kind invitation.

7. References

[1] A. Krasznahorkay *et al.*, *Phys. Rev. Lett.*, 116, 042501 (2016).
[2] F. Hug *et al.* LINAC16 Conference Proceedings MOP106012, pp. 313–315 (2017).
[3] D. Becker *et al.*, *Eur. Phys. J. A*, **54**, 11, 208 (2018).
[4] T. Stengler *et al.*, Proceedings of SRF2015, Whistler, BC, Canada (2015).
[5] J.D. Bjorken *et al.*, *Phys. Rev. D* **80**, 075018 (2009).
[6] E. Izaguirre *et al.*, *Phys. Rev. Lett.* **115**, 251301 (2015).
[7] I. Peric *et al.*, *Nucl. Instrum. Methods* **A 731**, 131 (2013).
[8] S. Grieser *et al. Nucl. Instrum. Methods* **A 906**, 120-126 (2018).
[9] B.S. Schlimme *et al. Nucl. Instrum. Meth. Phys. Res.* A **1013**, 165668 (2021).
[10] H. Merkel *et al.*, Phys. Rev. Lett. **106**, 251802 (2011).
[11] H. Merkel *et al.*, *Phys. Rev. Lett.* **112**, 221802 (2014).
[12] J.C. Bernauer *et al. Phys. Rev. C* **90**, 015206 (2014).
[13] R. Pohl *et al. Nature*, 466, 213 (2010).
[14] I. Friŝčić *et al.*, Phys. Rev. C **100**, 025804 (2019)
[15] S. Lunkenheimer, PhD Thesis (2022), https://openscience.ub.uni-mainz.de/handle/20.500.12030/6819
[16] L. Doria *et al.*, Proceedings of CIPANP 2018, arXiv:1809.07168.
[17] M. Lauß *et al. Nucl. Instrum. Meth. Phys. Res.* A 1012, 165617 (2021).
[18] M. Christmann *et al.*, *PoS*, **EPS-HEP2021**, 129, (2022).
[19] M. Battaglieri *et al.*, submitted to JLab PAC 45, arXiv:1712.01518 (2017).
[20] J.B.R. Battat *et al.*, *Astroparticle Physics*, 91, 65-74 (2017).

Journal of Physics: Conference Series **2391** (2022) 012006 doi:10.1088/1742-6596/2391/1/012006

Operation of an Energy Recovery Linac with an Internal Target

S V Benson[1] **and D R Douglas**[2]

[1] Thomas Jefferson National Accelerator Facility, Newport News, VA 23606
[2] David R. Douglas Consulting, Yorktown VA 23690

E-mail: felman@jlab.org

Abstract. Because an Energy Recovery Linac (ERL) decelerates and dumps the beam after at most a few passes, beam degradation in the interaction region that is orders of magnitude larger than in a storage ring is tolerable in such a device. As a consequence, some new types of Nuclear Physics experiments can be carried out. In 2016, an experiment (the DarkLight experiment) was installed in the Jefferson Lab FEL Driver ERL to test out this idea. The ERL had previously been used for FEL applications where high peak current and a large growth in energy spread was present. For the internal target, the machine setup required very small energy spread, and had to deal with large transverse emittance growth in the target. Additionally, the addition of a strong solenoid in the transport complicated the details of energy recovery. This presentation will describe how these new machine physics challenges were addressed.

1. Introduction

Internal target experiments to date have generally been carried out in storage rings [1]. We describe here an attempt to use an Energy Recovery Linac (ERL) to provide an opportunity to increase the density of the internal target by several orders of magnitude. An ERL can tolerate beam losses of as much as 0.1% after the target if they are in a well-shielded dump. This means that the target may be much thicker than in a storage ring, where losses must be very small ($\ll 1$ ppm per pass).

Another advantage of an ERL for internal target experiments is the ability to operate at very low energies. Dark Matter experiments often operate at low center-of-mass energy to restrict the phase space of the interaction [2]. Storage rings suffer from intra-beam-scattering (IBS) at low energy, limiting available beam brightness [3]. An ERL can provide much smaller emittance and energy spread since the beam only traverses the interaction region beamline and target once.

The DarkLight experiment embedded a 0.77 m long dense gaseous hydrogen target and detector package in a 1.1 m solenoid operating with 5 kG field [4]. A series of baffles, with small (3 mm diameter) apertures and differential pumps isolated the target from the ERL transport system [5]. The spent beam was transported from the target/detector to the downstream recovery line through a "Møller dump" that absorbed large-amplitude Møller-scattered electrons and collimated them using rings of carbon and tungsten. The ERL transport system then had to be optimized to provide sufficient control of the beam to the target, adequate downstream acceptance (both transverse and longitudinal) to handle the beam after interaction with the target, and control of solenoid-induced coupling so as to manage both beamline optics and the beam breakup instability. This combination of constraints was most readily satisfied by locating the experiment in the infrared FEL region of the ERL [6], with the experimental package replacing the infrared wiggler and appropriate modifications of the upstream and downstream transport providing the necessary phase space management (see figure 1-3).

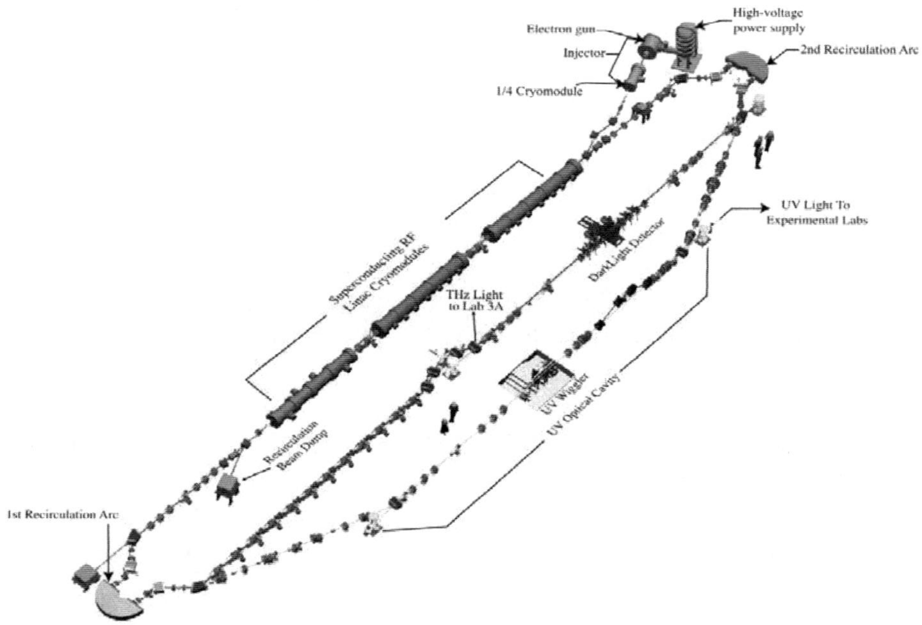

Figure 1. Jefferson Lab Energy Recovery Linac with DarkLight experiment embedded in the former Infrared FEL location.

Figure 2. Detail of the interaction region of the DarkLight experiment showing the 12 regular and 10 skew quadrupoles used to match into and out of the solenoid (blue cube) and rotate the transverse phase space to reduce the Beam Breakup instability.

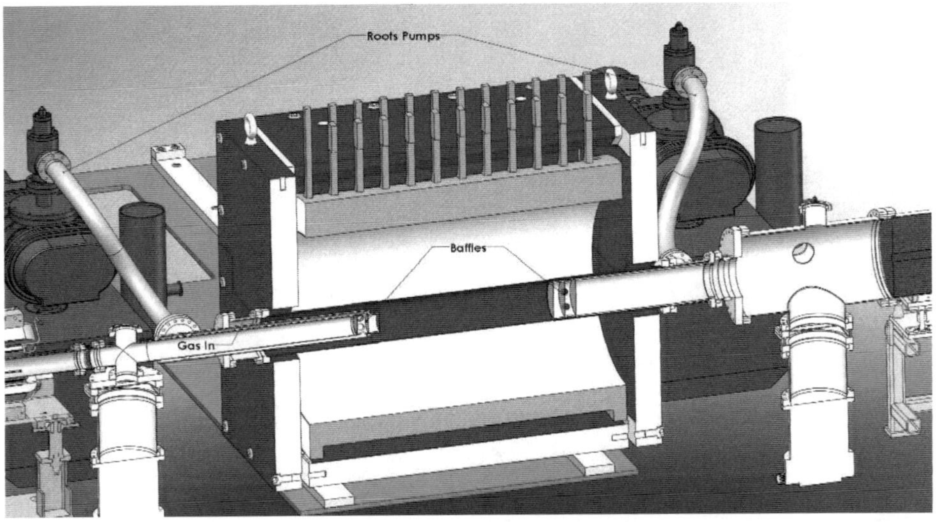

Figure 3. Detail of the interaction region of the DarkLight experiment. The Møller dump is on the far right and has a 1 cm graphite and tungsten aperture to capture scattered electrons.

2. Requirements

Ideally, the ERL transport system together with the target package should meet a simple set of requirements:

(i) The longitudinal match will provide a long bunch with small momentum spread to the target. This minimizes wake and resistive wall heating and provides narrow energy bandwidth for the user. The exhaust bunch will have a slightly degraded energy spread due to interaction with the target.

(ii) The target will be "thin" – in the sense that little betatron motion occurs over its length.

(iii) To avoid scattering-induced coupling, the beam will be uncoupled at the target. Thus, the linac to solenoid-center transport will be uncoupled and the solenoid-center to linac transport will be uncoupled.

(iv) To suppress Beam BreakUp (BBU), the linac-to-linac transport will exchange transverse phase spaces. This exchange is accumulated in the match to the target, the solenoid, and the match from the target.

(v) The beam will be matched into the solenoid with an upright phase space ($\alpha_{Twiss} = 0$) at the characteristic Twiss envelope value for a solenoid (twice the Larmor radius).

(vi) With the beam uncoupled at the "short" target, scattering will notionally not introduce coupling, so the target will transform the transverse phase spaces in a manner discussed elsewhere [7] and the beam from the target forward is characterized by an emittance increased - and a Twiss beta decreased - by the ratio of the incoming to outgoing angular divergences. Downstream betatron acceptance will be set to accommodate this exhaust phase space.

We note that an electron scattered within the solenoid will undergo motion characteristic of a magnetized beam. However, instead of a linear combination of the desirable "magnetized" eigenmodes, which follow solenoid field lines, it instead enters into a linear combination of the orthogonal "Larmor" eigenmodes (which are "magnetized" eigenmodes for a solenoid of opposite polarity) and precesses around field lines.

Table 1. Beam and Machine Parameters for DarkLight experiment.

Parameter (Unit)	Value
e- Beam Momentum (MeV/c)	100
Normalized Emittance (mm-mrad)	
upstream	8
downstream	~ 40
RMS Bunch Length (psec)	2.5
RMS Full-Energy $\frac{\delta p}{p}$(%)	
upstream	0.03
downstream	0.1 [4]
Bunch charge (pC)	60
Rep rate (MHz)	75
Current (mA)	4.5
Solenoid Field (kG)	5
Solenoid Length (m)	1.1
Solenoid Matched	
$\beta(m)$	1.3343
$\alpha(rad)$	0
Solenoid Radial Aperture (m)	0.1
Target Length (m)	0.77

In Table 1 we summarize the specifications for the accelerator and target for the DarkLight experiment for the initial runs. The average current and repetition rate was chosen only because it was straightforward to achieve. Future experiments would benefit from a higher repetition rate of 150 MHz and higher average current of at least 9 mA.

3. Beam Matching

The beam/target interaction has specific consequences for the beam [8]:

(i) Single scattering dominates;

(ii) The target is "thick" – it acts on the beam over its full length of 0.77 m.

(iii) Scattering events result in losses of order 200-300 ppm in the downstream Møller dump (which has 1 cm diameter aperture and is 0.5 m full length).

(iv) Scattering events increase the beam energy spread from the initial 0.03% rms to 0.1% rms.

Taking these issues into account, the following design approaches were taken:

The solenoid and target were, in simulations, subdivided into multiple slices of 0.055 m length. Twenty slices were used to describe the solenoid. Single scattering was modeled at fifteen points between 16 of these slices by introducing a pair of transformations eliminating the fringe field of the up- and downstream slice solenoid maps [7], and embedding between the "antifringes" a transformation inflating the beam angular divergences and momentums spread by user-defined factors. Given the (assumed) axial symmetry of the optics, this scattering transformation was therefore of the following form:

$$\begin{pmatrix} 1 & 0 & 0 & 0 & 0 & 0 \\ 0 & F_{\text{transverse}} & 0 & 0 & 0 & 0 \\ 0 & 0 & 1 & 0 & 0 & 0 \\ 0 & 0 & 0 & F_{\text{transverse}} & 0 & 0 \\ 0 & 0 & 0 & 0 & 1 & 0 \\ 0 & 0 & 0 & 0 & 0 & F_{\text{longitudinal}} \end{pmatrix}$$

We note that this transformation is not symplectic, which is however consistent with the notion of scattering being a nonconservative process when viewed solely from the perspective of the beam. We also note that the anti-fringe transforms [7] are not symplectic when used with variables $x, x', y,$ and y'. This had consequences for the particular coupling management scheme that was used.

In practice, ray-tracing was used to establish working values for $F_{\text{tranverse}}$ and $F_{\text{longitudinal}}$. An initial beam distribution was defined and traced through the solenoid to the Møller dump, with the F factors changed at the first slice and the results recorded. This was repeated at each of the rest of the slices. The inflation factors were then increased until a) the integrated loss on the Møller dump from scattering at all slices was of order 200 to 300 ppm, and b) the output momentum spread from each (and all) slice(s) was 0.1% rms. The resulting distributions were then notionally consistent with those produced by the analysis of Ref. [8], and give rms beam divergences and energy spread at the solenoid exit of the order of earlier analytic estimates.

This analysis was for an ideal system. We had, however, to take into account some realities. First, the target was not short – it was long enough for electrons to respond in position to deflections from scattering. Secondly, the dynamics within a solenoid must be observant of the effects – or rather, the absence – of the solenoid fringe fields when the beam is scattered inside the solenoid [9] This had numerous consequences for the modeling of the system using conventional beam optics codes such as DIMAD. In particular, the solenoid maps in such programs inherently include fringe effects, so when subdividing the solenoid, provision for eliminating fringes in the interior must be made as described above. Given that in an (x,x',y,y') representation of the transverse phase space these descriptions are nonsymplectic, it was thus not possible to manage coupling compensation in the manner demanded by the above requirements.

As noted above, scattering in the solenoid interior put electron orbits into a linear combination of magnetized and Larmor modes [9]. The beam was therefore inherently coupled. Given that the transform from the interior of the solenoid back to the linac was nonsymplectic (by virtue of the anti-fringe fields in the solenoid), it was not possible to compensate this using solely a downstream quadrupole telescope (which can only produce symplectic transformations).

It was thus not possible to "orthogonalize" the optics solution; system requirements therefore had to introduce opportunity to optimally manage coupling (so as to control BBU) while still providing control over beam envelopes through the constrained apertures of the target and during energy recovery. The following set of requirements was found to allow this.

3.1. Accommodations for real-life effects

(i) The target was "thick": motion of electrons scattered in the front end of the target evolved significantly under the influence of the solenoid field. An analytic description is given in [8].

(ii) The beam – which was uncoupled at the front end of the interaction region – may be coupled at any point thereafter and *will* be coupled throughout the length of the target and solenoid. scattering may enhance the coupling. The recovery optics solution must be optimized to transport the degraded coupled beam to the dump.

(iii) Matching the beam to the solenoid entrance will – even for coupled beam – prevent beam envelope mismatch and limit overall beam size in the solenoid, providing optimal transmission at the target aperture restrictions.

(iv) As the beam was coupled and the target was thick, scattering enhanced coupling and degraded emittance. However, the degradation – described in [8] – resulted in an exhaust beam with well-defined phase space characteristics, with moments (sigma matrix elements, including size and divergence) as well as specific patterns of coupling and distortion that inform optimization of the recovery betatron match. Downstream betatron acceptance was therefore tuned to accommodate this exhaust phase space.

(v) The Møller dump served to provide some degree of collimation of the exhaust electron beam. It was downstream of the detector solenoid, immediately upstream of the first quadrupole downstream of the solenoid, of length 0.5 m and 5 mm radial bore (1 cm diameter). Most of the beam clears the dump (see figure 4).

4. Accelerator Setup

Several accomplishments were required to carry out the DarkLight experiment:

(i) Clean beam had to be transported through six 3 mm holes in the Kapton foils. To ensure that we could accomplish this, we ran an aperture test to verify that we could run 500 kW of beam through a very small hole for hours at a time [10, 11].

(ii) We had to run with very small energy spread with full energy recovery. This involved accelerating off crest in first and third cryomodules and off-crest in the opposite direction in the middle cryomodule producing a smaller overall energy spread that could be quickly phased. The recovered beam was 180° out of phase with the accelerated beam. This longitudinal phase space match was tested out in the aperture test and worked well.

(iii) Scattering upstream of the target had to be minimized. This was done by careful matching of the beam and running the accelerator cavities at $\sim$ 6MV/m gradient so there would be no field emission.

(iv) We had to transport a low energy beam through a large 5 kG solenoid. This is straightforward if the solenoid is ideal. In our case there was an uncompensated conductor in the solenoid that produced a transverse field in the solenoid.

(v) As noted above, the phase space had to be rotated by 90 degrees to increase the BBU threshold. This is partly accomplished by the solenoid field but it is completed by 5 skew quadrupoles that rotate the beam by about 22 degrees before and after the solenoid. Several beam position monitors and viewers had to be added to the transport line to set up these skew quadrupoles;

The match to the solenoid (upstream telescope configuration) was developed as follows. The skew quadrupoles were set to provide a phase space rotation of 22° from the linac to the solenoid entrance. The match optimization was performed by putting a decoupling +45° roll at the center of the solenoid, and adjusting the quadrupoles to zero out the off-diagonal elements of the linac-to-solenoid-center transfer matrix ($M_{13} = M_{14} = M_{23} = M_{24} = M_{31} = M_{32} = M_{41} = M_{42} = 0$) while matching the beam envelopes to the acceptance of the solenoid. The latter constraint was most simply applied by injecting a matched decoupled beam of unit (1 m-rad) emittance into the start of the recirculator, and fitting the sigma matrix elements to the solenoid acceptance envelopes: $\Sigma_{11} = \beta, \Sigma_{12} = 0, \Sigma_{21} = 1/\beta, \Sigma_{33} = \beta, \Sigma_{34} = 0$, and $\Sigma_{44} = 1/\beta$. There were eight independent constraints on the eleven free parameters.

The match from the solenoid was rendered somewhat more complex by virtue of the need to manage beam that had been scattered by the target. The design/optimization process was therefore iterative The impact of scattering was initially treated as an impulsive increase the in the angular divergence of an upright matched beam at the very center of the solenoid. The downstream lattice was then matched to accept a beam with emittance scaled up by a

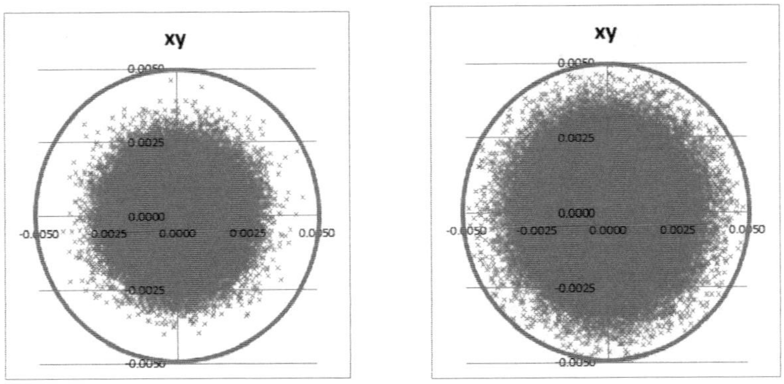

Figure 4. Beam profiles at the graphite entrance of the Møller dump (left) and at the Tungsten exit (right). The red circle represents the 1 cm diameter hole in the dump.

degradation factor, and beam envelope reduced by the same factor. In this case, the estimated emittance blowup was approximately a factor of four, so an upright beam ($\alpha = 0$) with envelope value of $\sim (1/4)\beta_{\text{solenoid}}$ was transported from the center through to the re-injection point of the linac, and two stages of matching were performed. The first stage set the 11 quads immediately downstream of the solenoid; these were optimized to complete the phase space exchange and to generate an optimized lattice envelope match at the center of the return transport arc. As the phase space exchange is complete by the end of the target insertion region, the remaining transport is decoupled and described by the usual lattice parameters. In practice, these constraints were – as with the match to the solenoid – entered via the sigma matrix, in as much as the initial beam specification and upstream lattice are fully coupled. Again, the eleven quads are under-constrained by the eight independent target parameters, so some latitude was available in the optimization process; internal mismatch and aberrations were thereby afforded some control. The match for this setup is shown in figure 5. The beam was – in simulation – very well contained throughout the beampipe to the dump.

We then had to take into account the thickness of the target. The degradation factor $F_{\text{transverse}}$ was set by comparison to losses as evaluated in [8]; that is, it was adjusted to give 200-300 ppm loss in the Møller dump. $F_{\text{longitudinal}}$ is set to give 0.1% rms dp/p by the end of the solenoid.

Once the exhaust beam was thus defined, full optimization of the recovery match proceeded. The initial solution was used to propagate beam matrices and as the basis for raytracing. Beam matrices for scattering from each "slice" of the target were propagated and spot sizes evaluated. Particle losses and distributions at key locations – solenoid exit, linac reinjection point, and at the end of the linac – were generated and cross-checked with the beam matrix propagation. The reinjection match was adjusted to provide optimum transmission (that is, to minimize the envelope of the collection of all propagated spot sizes). The results of the final match are shown in figure 6. Though increased from the thin target, the overall beam size is still well confined.

Longitudinal phase space behavior was monitored during the matching process to ensure that the appropriate induced momentum spread was adequately managed during recovery. The impact of chromatic aberrations was evaluated to confirm that the matching process did not lead to excessive variation in relevant beam parameters.

5. Accelerator Performance

An attempt was made to energy recover beam with no hydrogen gas or foils. A reasonable match was achieved but CW beam was not run before adding the target foils due to lack of time.

The foils and hydrogen system were installed and we verified that up to 2 Torr of hydrogen

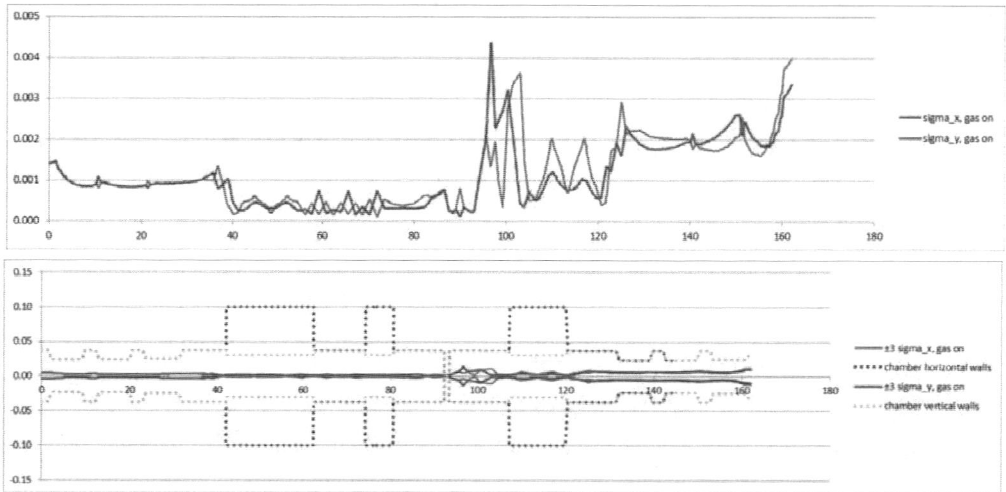

Figure 5. RMS beam envelopes for a thin target (top) and three times the RMS envelope compared to the beampipe aperture (bottom). All units are in meters.

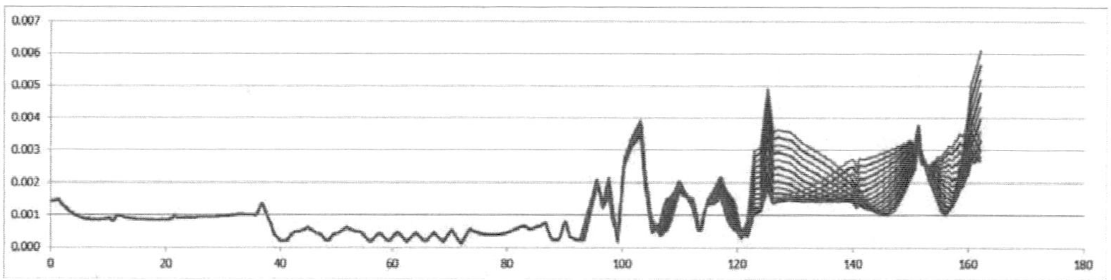

Figure 6. Beam envelopes for a thick target rematched to the dump. Compare to figure 5.

could be injected into the target while maintaining ion pumps at the end of the target region.

When we tried to put beam through the target however, we found two problems:

(i) The transverse field in the solenoid created an arcing trajectory through the target. It was possible to get the beam to go through the upstream and downstream sets of foils but the trajectory was not straight;

(ii) We found that one of the foils was mis-aligned by about 1 mm. This was enough to reduce the aperture to about 2 mm, which was too small for the electron beam to be transported cleanly through the target.

Though we could run pulsed beam with significant losses in the foils, we would not run CW beam.

6. Conclusions

The basic setup of an internal target was tested. It is a complicated experiment and requires significant commissioning time, which was not available. It is very important to test out the full beam transport at full current without foils or gas before target installation. This was not done. Once the beam transport is properly configured, the target can be installed and the beam transported, first without gas and then with gas. Once clean transmission has been achieved, the current can be ramped up to its full value.

6.1. *Acknowledgments*

This material is based upon work supported by the U.S. Department of Energy, Office of Science, Office of Nuclear Physics under contract DE-AC05-06OR23177.

References

[1] D. Hasell et al. "The BLAST Experiment" 2009 Nucl. Instl and Meth. A603 247 https://doi.org/10.1016/j.nima.2009.01.131, and
R. J. Holt, "Nuclear Physics with internal targets in electron storage rings" AIP Conference Proceedings **161**, 138 (1987)
https://doi.org/10.1063/1.36883.
[2] P. Ilten et al., "Experiments and Facilities for Accelerator-Based Dark Sector Searches" arXiv:2206.04220 [hep-ex] https://doi.org/10.48550/arXiv.2206.04220
[3] A. G. Ruggiero, "Intrabeam scattering in electron and proton storage rings (a review)", AIP Conference Proceedings **123**, 424 (1984)
https://doi.org/10.1063/1.34887.
[4] J. Balewski et al., "The DarkLight Experiment: A Precision Search for New Physics at Low Energies", arXiv:1412.4717 [physics.ins-det]
https://doi.org/10.48550/arXiv.1412.4717.
[5] S. Lee et al., "Design and operation of a windowless gas target internal to a solenoidal magnet for use with a megawatt electron beam", Nucl. Instr. and Meth. , **A 939** (2019) 46–54
https://doi.org/10.48550/arXiv.1903.02648
[6] C. Tennant "LERF – New Life for the Jefferson Laboratory FEL", Proceedings of ERL 2017, TUIACC001
http://jacow.org/erl2017/papers/tuiacc001.pdf
[7] D. Douglas "Beam Optics for Transverse Scattering Management In An ERL", JLAB-TN-12-046, 26 September 2012.
[8] C. Tschalaer, "Target Scattering Phase Space", Bates-internal report B/IR-16-01, 28 June 2016.
[9] D. Douglas, S. Benson, and C. Tennant, "A Simple (and Quite Possibly Incorrect) Description of Magnetized Beam Formation (That Gives the Right Answer Anyway...)" JLAB-TN-18-006,
D. Douglas, "Betatron Matching Magnetized Beams", JLAB-TN-15-026, and
D. Douglas, "Electron Optics for Cooling Insertions", JLAB-TN-16-046.
[10] D. Douglas "IR FEL Driver ERL Configuration for the DarkLight Aperture Test", JLAB-TN-13-009, 11 December 2012.
[11] R. Alarcon et al. "Transmission of Megawatt Relativistic Electron Beams Through Millimeter Apertures", Phys. Rev. Lett. 111 (2013) 164801
https://doi.org/10.48550/arXiv.1305.0199

A new light particle is being born

A.J. Krasznahorkay[1], M. Begala[1,3], M. Csatlós[1], L. Csige[1], J. Gulyás[1], A. Krakó[1,3], A. Krasznahorkay[1] [1], B.M. Nyakó[1], I. Rajta[1], N.J. Sas[1,3], J. Timár[1], I. Vajda[1]

[1]Institute for Nuclear Research (ATOMKI), H-4001 Debrecen, P.O. Box 51
[3]University of Debrecen, 4010 Debrecen, PO Box 105, Hungary

E-mail: `kraszna@atomki.hu`

Abstract.
A few years ago we observed anomalous electron-positron angular correlations for the 18.15 MeV M1 transition of ^{8}Be. This was interpreted as the creation and decay of an intermediate bosonic particle with a mass of m_0c^2=16.70±0.35(stat)±0.5(sys) MeV, which is now called X17. The possible relation of the X17 boson to the dark matter problem triggered an enormous interest in the wider physics community. We also observed a similar anomaly in ^{4}He, which could be described also by the creation and subsequent decay of the same X17 particle. Very recently, the ^{11}B proton capture reaction was used for exciting the 17.2 MeV broad (Γ= 1.15 MeV) resonance in ^{12}C and studying their internal pair creation decay. Anomalies were observed in the angular correlation of the electron-positron pairs, which together with the ^{8}Be and ^{4}He data provides kinematic evidence for the X17 particle and supports their vector boson and fifth force explanation.

1. Introduction

Our current knowledge of nature at the fundamental level is successfully explained by the Standard Model (SM) of elementary particles, but this theory describes only five percent of the entire content of the Universe. We named the overwhelming unknown constituents as dark matter and dark energy. The nature of dark matter is currently one of the greatest unsolved mysteries in physics.

The search for new physics beyond the SM can be divided into two categories. The first one is the search for new heavy particles and interactions at high energies, the so-called "energy frontier" research. A complementary and vital role is played by low-energy, precision and/or high-intensity experiments, which require a joint effort of the particle, atomic, and nuclear physics communities [1, 2]. One of the main targets of the intensity frontier in particle physics is a new force carrier which is much lighter than the weak scale and which interacts with the Standard Model particles very weakly. Observations of rare nuclear transitions can be used to search for new hidden force-carrier particles at the MeV scale. In this approach, a fixed target is bombarded with a hadron beam to produce excited states of a nucleus. The excited state then also decays by emitting new particles.

Recently, we have observed an anomaly in the nuclear decay of ^{8}Be [3]. The ^{7}Li(p,γ)^{8}Be reaction was used to populate the excited states in ^{8}Be selectively and the differential internal

[1] Present address: CERN, Geneva, Switzerland

pair conversion coefficients were studied for the 17.6 MeV, and 18.15 MeV ($J^\pi = 1^+ \rightarrow 0^+$) M1 transitions in ^{8}Be. Significant, peak-like enhancement of the internal pair creation [4] was observed at large angles in the angular correlation of the 18.15 MeV transition, but not in the 17.6 MeV one [3]. This observation was interpreted as the creation and the subsequent decay of an X boson with mass m_0c^2=16.70±0.35(stat)±0.5(sys) MeV. The branching ratio of the e^+e^- decay of such a boson to the γ decay of the 18.15 MeV level of ^{8}Be is found to be 5.8×10^{-6} [3].

J. Feng and his group [5, 6] studied our data along with other previous experiments and showed that our observation strongly disfavors dark photons. Instead, they proposed a new theory and explained the observation by introducing a fifth fundamental force [5]. If confirmed by further experiments, this discovery of a possible fifth force would completely change our understanding of the universe, by taking a step towards the unification of forces and dark matter.

The possible relation of the X boson to the dark matter problem and the fact that it might explain the $(g\text{-}2)_\mu$ puzzle triggered an enhanced theoretical and experimental interest in the particle and hadron physics community [7]. Inspired by this enhanced interest, we studied the effect in other nuclear reactions using improved experimental conditions.

2. Measurement of e^+e^- angular correlation

The high-lying energy levels of ^{8}Be were excited by proton radiative capture on ^{7}Li target. The protons were produced using a home-built Van de Graaff accelerator at ATOMKI. In order to measure the angular correlation of the e^+e^- pairs, we have built a highly efficient spectrometer with high angular resolution that no one has used before [8]. We are talking about observing a very rare process, the probability of which is one million times (10^{-6}) lower compared to the probability of electromagnetic transition (γ-radiation). The detection of the new particle is further complicated by the fact that in high-energy electromagnetic transitions, according to quantum electrodynamics, e^+e^- pairs can also be created. These provide a significant background for our measurements, since the probability of the process called Internal Pair Creation (IPC) is about one thousandth (10^{-3}) of the probability of a γ-transition. However, the angular correlation between the resulting e^+e^- pairs is well known, and with good approximation the IPC rate decreases exponentially depending on the angle between the two particles, and to contrast that to the sharp peak around a non-rero opening angle for the decay of a massive particle. Detecting these very rare events was a significant challenge for our e^+e^- coincidence spectrometer. The spectrometer consists of 5 Multiwire Proportional Counters (MWPC) and thin (ΔE) and thick (E) plastic scintillation detectors. The schematic diagram of the spectrometer can be seen in Figure 1.

The detectors of the spectrometer were sufficiently stable during the typically 1-week experiments, we only saw changes in the gain of a few % in the scintillation detectors, which were corrected during the data analysis. The efficiency (acceptance) of the spectrometer was determined experimentally, depending on the correlation angle, using uncorrelated e^+e^- events collected simultaneously with e^+e^- coincidence events. With this efficiency, we corrected the angular correlation of all internal pairs coming from the target. In Figure 2(a), the energy-sum spectrum of e^+e^- pairs measured after the decay of the 17.6 MeV excited state of ^{8}Be is presented.

This state was excited by resonant proton capture. The energy of the resonance is E_p= 441 keV, and the width is Γ=10.7 keV. The 17.6 MeV transition in the spectrum goes to the ground state of ^{8}Be, while the 14.6 MeV goes to the very broad first excited state, decaying to two α-particles. The intense ^{16}O line at the beginning of the spectrum is excited by the ^{19}F(p,α)^{16}O reaction in the LiF target used in this case. In Figure 2(b) the angular correlation of the 17.6 MeV and 14.6 MeV transitions are indicated. The full curves marked with M1 and E1 are the result of simulations assuming magnetic and electrical dipole transitions.

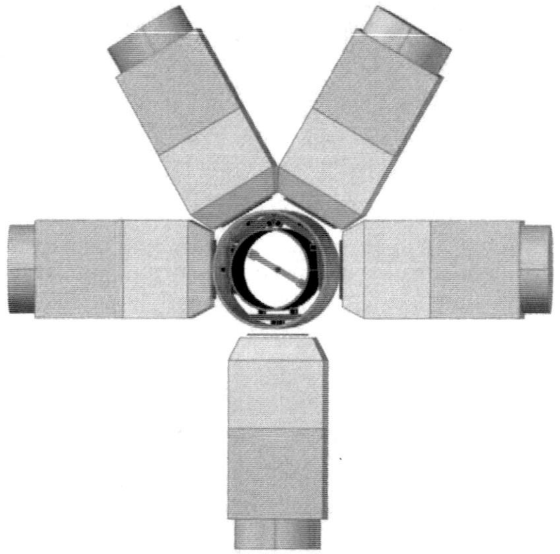

Figure 1. Schematic diagram of the electron-positron pair spectrometer.

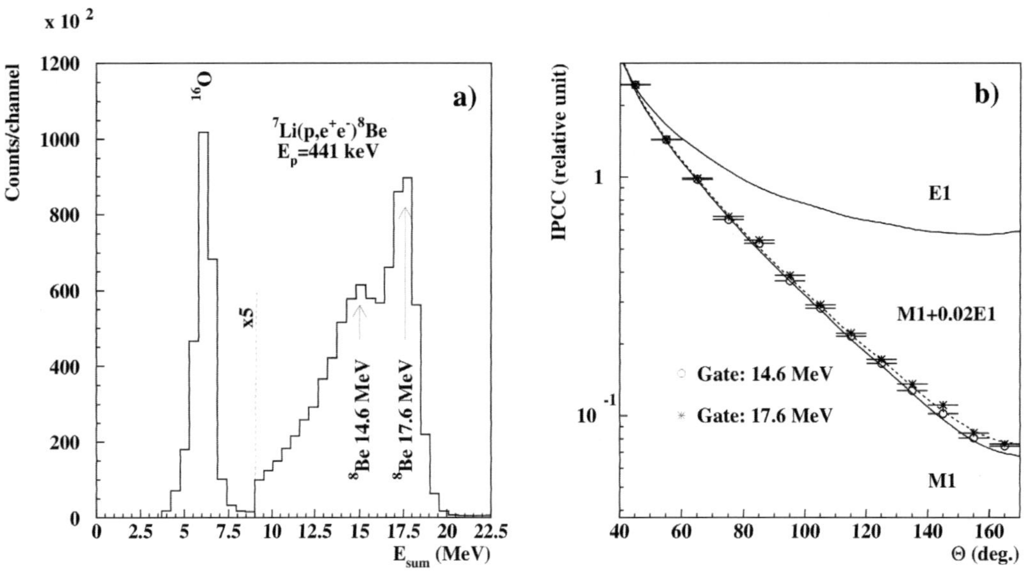

Figure 2. Measured total energy spectrum (a) and angular correlation (b) of the e^+e^- pairs originated from the decay of the 17.6 MeV resonance compared with the simulated angular correlations assuming M1 (full curve) and M1+1.4%E1 mixed transitions (dashed line).

Monte Carlo simulations for the experiment were conducted with the GEANT [9] code developed at CERN. Every part of the spectrometer, to the last screw in the vicinity of the target, is carefully incorporated into the simulation to obtain the spectrometer's response to both e^+e^- pairs and intense γ radiation. In addition to the internal pair creation process, we also took into account the background caused by γ radiations, external pair creation and e^- and

e^+ multiple scattering in order to understand as accurately as possible the response functions of the detectors and spectrometers.

As shown in Figure 2, in line with previous results, above 110° we have observed a slight difference between the experimental values and the values simulated for the M1 transition. As far as we know, this transition should be a clean M1 transition. However, if we also take into account the effect of the weak, non-resonant E1 multipolarity background generated by direct proton capture by mixing only 1.4% E1 angular correlation to the M1 angular correlation, then we can already interpret our experimental results accurately. The degree of continuous background mixing to the resonance depends on both the width of the resonance and the thickness of the target, which causes the broadening of the resonance.

After this experiment, we also examined the angular correlation of e^+e^- pairs from a higher excited state of 18.15 MeV, which was also well known. Our experimental angular correlations measured on the E*= 18.15 MeV resonance can be seen in Figure 3.

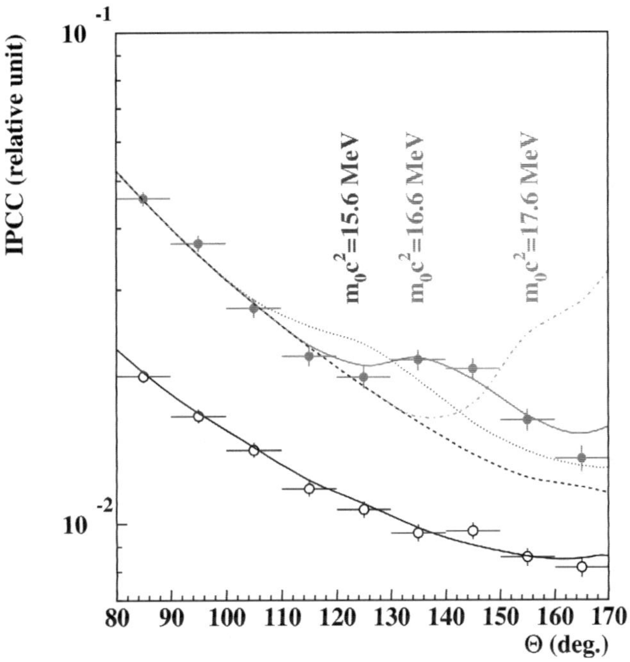

Figure 3. Experimental angular e^+e^- pair correlations measured in the ^{7}Li(p,e^+e^-)^{8}Be reaction at E_p=1.10 MeV with -0.5 $\leq$ y $\leq$ 0.5 (full circles) and $|y| \geq 0.5$ (open circles) (see the definition of y in the text). The results of simulations of boson decay pairs added to those of IPC pairs are shown for different boson masses, with different colors.

This state can also be excited resonantly, with protons of 1.040 MeV, but the resonance is much wider (Γ=138 keV) than the previous resonance. Therefore, a much larger mixing can be expected from the non-resonant E1 background. This was confirmed by previous experimental results published for the range of 75° to 130° [10].

In this experiment, we extended our angular range up to 170°, and as a result of the ≈3 orders of magnitude larger efficiency compared to the spectrometer built by Stiebing et al.[10], we also significantly reduced the statistical error of measurement.

It is important to note that the distributions were collected for e^+e^- pairs with a near symmetrical energy distribution:

$$-0.5 < \frac{E_{e^-} - E_{e^+}}{E_{e^-} + E_{e^+}} < 0.5 \ , \quad (1)$$

where E_{e^-} represents the energy of the electron, while E_{e^+} represents the energy of the positron. To the best of our knowledge, the significant anomaly observed at 140° cannot currently be explained reliably by a nuclear physics effect. The difference between the experimental and theoretical values is significant and can be explained by the introduction of a new particle (X17) with a mass of 16.7±0.35 (statistical error) ±0.5 (systematic error) MeV. The probability of formation of e^+e^- pairs when the particle decayed was 5.8×10^{-6} compared to the probability of a γ transition (branching ratio).

Assuming X17, the different theoretical descriptions give different values for the X17 branching ratios of the 17.6 MeV excited state as well [5, 6, 11, 12]. This should be further investigated both experimentally and theoretically.

3. Experiments with the new Tandetron accelerator

3.1. Study of the e^+e^- angular correlation from the $^3H(p,e^+e^-)^4He$ reaction [13]

The e^+e^- angular correlation was studied from the above reaction at E_p = 510, 610 and 900 keV proton energies to search for signatures of the X17 particle. This reaction induced direct capture [14] and resonant capture and populate the overlapping $J^\pi = 0^+$ first, and $J^\pi = 0^-$ second excited states in ^{4}He [15].

The proton beam with a typical current of 1.0 μA was impinged on a ^{3}H target for about 100 hours for each bombarding energy. The ^{3}H was absorbed in a 4.2 mg/cm^2 thick Ti layer evaporated onto an 0.4 mm thick molybdenum disk with a diameter of 50 mm. The density of the ^{3}H atoms was $\approx 2.7 \times 10^{20}$ atoms/cm^2. The disk was cooled down to liquid N_2 temperature to prevent ^{3}H evaporation. In such a thick ^{3}H+Ti target, the proton beam was stopped completely. The target was shifted off the center of the spectrometer by 25 mm downstream along the proton beam axis to avoid the screening by the target backing and holder. We have used a large (50 mm diameter) cooling pipe with a fixing screw, which limited the maximal correlation angle to $\approx 140°$, since the surface of the target was inside the pipe.

The proton energies were chosen to stay below the threshold of the (p,n) reaction (E_{thr}=1.018 MeV). The ^{4}He nucleus was excited up to E_x = 20.21, 20.29 and 20.49 MeV at the used proton beam energies, so it was expected that the first excited state of ^{4}He ($J^p=0^+$, E_x=20.21 MeV, Γ=0.50 MeV) and the second one ($J^p=0^-$, E_x=21.01 MeV, Γ=0.84 MeV) were both populated [15].

Our previous experimental setup [3, 8] has recently been upgraded by the replacement of the scintillators with EJ200 ones and PM tubes by Hamamatsu 10233-100 ones. The sizes of the scintillators were $82 \times 86 \times 80$ mm^3 each.

As another improvement, the Multi Wire Proportional Chambers (MWPC) have been replaced by novel Double-sided Silicon Strip Detectors (DSSD), placed very close to the front face of the scintillators, to enhance the efficiency of the experimental setup and its homogeneity.

We also increased the number of telescopes from 5 to 6. The positions of the hits were registered by the DSSDs having sizes of 50×50 mm^2, strip widths of 3 mm and a thickness of 500 μm. The telescope detectors were perpendicular to the beam direction, each at 60° to its neighbors, around a vacuum chamber made of a carbon fiber tube with a wall thickness of 1 mm.

3.2. Validating the results of the simulations

In order to test the accuracy of the GEANT simulations for describing our experiments, we made measurements with the previously described ^{7}Li(p,γ)^{8}Be reaction.

A ^{7}Li target was evaporated onto the same Ti+Mo backing that the ^{3}H target used, in the same geometrical setup as described earlier. Since the internal pair creation coefficient calculated by Viviani et al. [16] for ^{4}He is only slightly higher than that calculated one for the 17.6 MeV transition in ^{8}Be, this transition allowed us to accurately test the backgrounds coming from γ radiation in our main experimental setup as well.

The ratio of the event numbers used for the simulations are determined by the internal pair creation coefficient as described before. These IPC+EPC simulations provided an accurate description of the experimental data. The contribution of the IPC and EPC processes to the events selected by our analysis in this case is about the same. These results convinced us that our spectrometer's behavior could be accurately described by our simulations, even in the presence of intense γ-rays and a thick target backing.

4. Experimental results

In order to search for the assumed X17 particle, both the energy-sum spectrum of the e^+e^- pairs measured by the telescopes, and their angular correlations, determined by the DSSD detectors, have been analyzed. For the real "signal" events we always required that the energy-sum for the e^+e^--pairs should be equal to the transition energy, which we want to investigate.

Since the counting rates in the detectors were low ($\approx$ 150 Hz in the scintillators and $\approx$ 25 Hz in the DSSD detectors) and the coincidence time window was sharp ($\approx$ 10 ns), the effect of random coincidences was negligible. In the following, we show only the real-coincidence gated spectra.

The down-scaled (x0.08) energy-sum spectrum of the e^+e^- pairs collected by all combinations of the telescope pairs in the 10 - 25 MeV energy range is shown by a dashed-line histogram in Fig. 4a) after subtracting the cosmic-ray background (CRB). This background was measured for two weeks before and after the experiments using the same gates and conditions as used for the in-beam data. It has been found that above E(sum)=25 MeV only the CRB contributes to the spectrum.

In order to reduce the External Pair Creation (EPC) background, we constructed a spectrum also from e^+e^- pairs, which were detected by telescope pairs with relative angles of 120°. The spectrum is shown in Fig. 4a) as a full-line histogram. The peak in the spectrum may come from the internal pairs created in the direct proton capture process or in the $0^+ \rightarrow 0^+$ E0 transition of ^{4}He and may also come from the e^+e^--decay of the X17 hypothetical particle. The background in the spectrum below E(sum)=17 MeV is created by external pairs induced by the γ-rays coming from the direct proton capture on the ^{3}H target.

The angular correlation spectra obtained for ^{4}He are indicated in Fig. 5(a) by dots, stars and full circles for E_p= 510, 610 and 900 keV, respectively. For better readability, the spectra are shifted by 1-1 orders of magnitude according to the labels.

The angular correlations of the e^+e^- pairs for the background region marked in Fig. 4 were also calculated, and compared to combinations of Monte Carlo simulations of different processes resulting in valid e^+e^- events in the spectrometer.

We found that the most significant background was provided by e^+e^- pairs created by γ-rays generated during direct proton capture on ^{3}H. For small correlation angles, this process can fully interpret the measured values. As a result of this, the normalization of the contribution of γ events was derived from the simulation's fit to the data in the 40° to 70° opening angle region.

Note that for the background regions of the sum-energy, the experimental and the corresponding simulated curves show a fairly good agreement over the entire angular range, thus validating the correctness of the simulations.

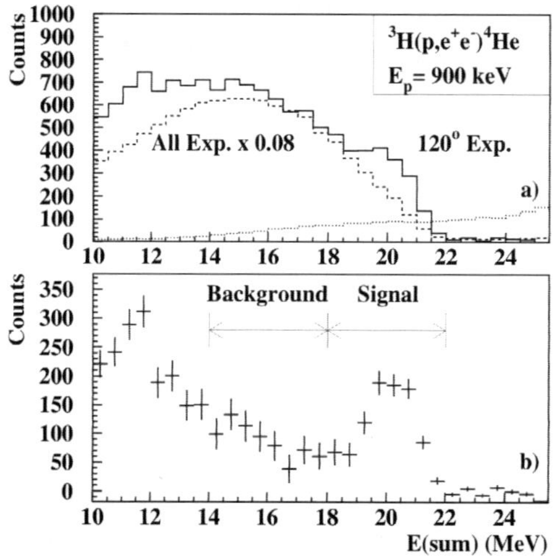

Figure 4. a): Experimental energy-sum spectra of the e^+e^- pairs derived, respectively, for "All" different detector combinations (dashed-line histogram with counts multiplied by 0.08) and for detectors at 120° relative angles (solid-line histogram). The cosmic-ray background contributions are subtracted from both spectra. The CRB spectrum corresponding to the "All" spectrum is plotted with dotted line. b): The markers with error bars show the difference of the solid-line and dashed-line distributions from Figure a).

Here we mention that in this case the usual method of background determination, i.e. performing the experiment without target material, cannot be applied because the main source of the background is the target material itself.

For further theoretical interpretation of the results shown in Fig. 5(a), the simulated angular correlations were subtracted from the experimental ones. The angular correlations (points with error bars) obtained after subtraction are shown in Fig. 5(b).

The corresponding proton beam energies are indicated in the figure. The anomaly previously observed and explained by the decay of the X17 particle appeared at each of the bombarding energies.

4.1. Fitting the angular correlations

In order to derive the exact value for the mass of the decaying particle from the present data, we carried out a fitting procedure for both the mass value and the amplitude of the observed peak.

The fit of the original experimental data was performed with RooFit [17] by describing the e^+e^- angular correlation with the following intensity function (INT):

$$INT = N_{EPC} * PDF(EPC) + N_{IPC} * PDF(IPC) + N_{Sig} * PDF(sig) \, , \qquad (2)$$

where $PDF(X)$ stands for the MC-simulated probability density function and N_X is the fitted number of the events of the given process. $PDF(sig)$ was simulated by GEANT3 incorporating the relativistic two-body decay of a particle with a given mass. Therefore, $PDF(sig)$ was constructed as a 2-dimensional model as a function of the e^+e^- opening angle and the mass of

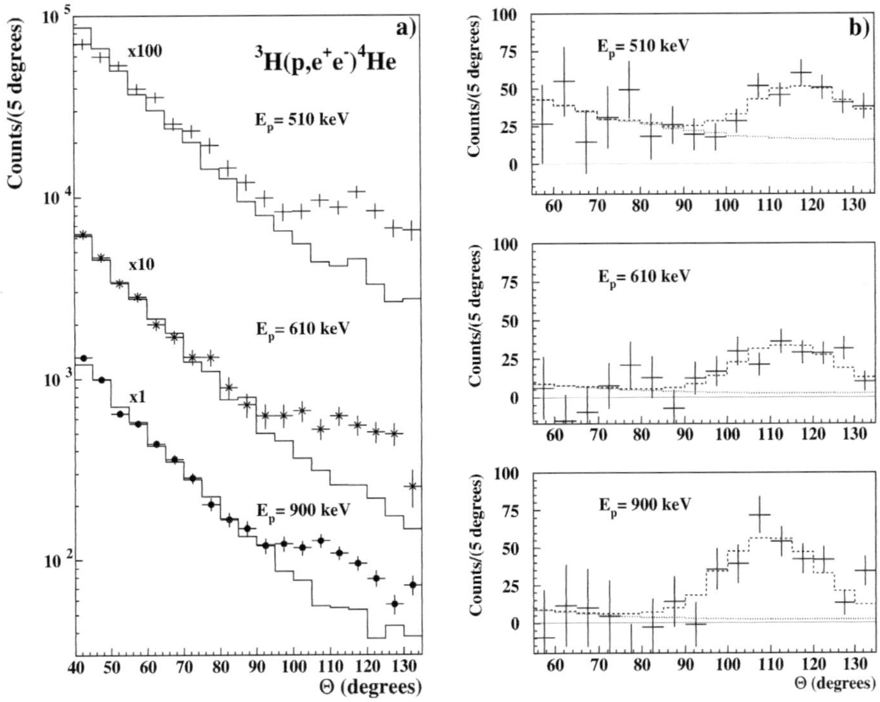

Figure 5. a): Angular correlations of the e^+e^- pairs for the "Signal" region (see Fig. 4). Symbols with error bars indicate experimental data measured in the ^{3}H(p,γ)^{4}He reaction at different proton beam energies, while solid-line histograms correspond to the respective data obtained in the simulations described in the text. b): Comparison of the experimental and the simulated angular correlations of the e^+e^- pairs. The best fitted sum (dashed line) as sum of the the simulated background (dotted line) and the simulated contribution of the hypothetical X17 boson is compared with the experimental signal values (dots with error bars)

the simulated particle. To construct the mass dependence, the PDF linearly interpolates the e^+e^- opening angle distributions simulated for discrete particle masses.

Using the intensity function described in Equation 2, we first performed a list of fits by fixing the simulated particle mass in the signal PDF to a certain value, and employing RooFit to estimate the best values for N_{Sig}, N_{EPC} and N_{IPC}. Allowing the particle mass to vary in the fit, the best fitted mass is calculated. We also made the fits by fixing the N_{EPC} value to the experimental data as described earlier.

The values obtained for the energy and branching ratio of the X17 boson and the IPC values, as a result of the average of the two fits described above, are summarized in Table 1, and the corresponding fits after subtraction of the EPC background are shown in Fig. 5(b).

4.2. Systematic uncertainties

The systematic uncertainties were estimated from the simulations. Taking into account the uncertainty of the target position along the beam line estimated to be ±2 mm, may cause an $\Delta m_X c^2 = \pm 0.06$ MeV uncertainty. The uncertainty of the position of the beam spot perpendicular to the beam axis was estimated to be ±2 mm in the worst case, which may cause a shift in the invariant mass of $\Delta m_X c^2 = \pm 0.15$ MeV. The total systematic error was

Table 1. Internal Pair Creation Coefficients (IPCC), X17 Boson branching ratios (B_x), masses of the X17 particle, and confidences derived from the fits.

E_p	IPCC	B_x	Mass	Confidence
(keV)	$\times 10^{-4}$	$\times 10^{-6}$	(MeV/c^2)	
510	2.5(3)	6.2(7)	17.01(12)	7.3σ
610	1.0(7)	4.1(6)	16.88(16)	6.6σ
900	1.1(11)	6.5(20)	16.68(30)	8.9σ
Averages		5.1(13)	16.94(12)	
^{8}Be values		6	16.70(35)	

conservatively estimated as: $\Delta m_X c^2(\text{syst.}) = \pm 0.21$ MeV.

4.3. Conclusions on X17

Table 1 shows the fitting parameters and the average of the parameters. As can be seen, consistent values were obtained for each fitting parameter. In the last row, our corresponding values measured in the case of ^{8}Be are also shown [3].

The obtained mass [13] agrees very well with that observed in the earlier ^{8}Be experiment, which is remarkable considering that the excesses in the observed angular correlation spectra appear at different correlation angles as one would indeed expect from the kinematics of the relativistic two-body decay. Therefore, our new observation enhances the possibility that the measured anomalies can be attributed to the same new particle X17.

As shown, the branching ratios of the X17 particle are identical within uncertainties, for the three beam energies proving that the X17 particle was most likely formed in direct proton capture, which has a dominant multipolarity of E1.

4.4. Preliminary results for the e^+e^- angular correlation from the $^{11}B(p,e^+e^-)^{12}C$ nuclear reaction.

The reaction was studied between $E_p = 1{,}5$ MeV and 2.9 MeV at 6 different bombarding energies. These energies covered the E_x= excitation energy region from 17.2 to 18.5 MeV, which included a wide resonance (E_x=17.23 MeV, J^π=1^-, Γ=1.05 MeV). The preliminary results are shown in Fig. 6.

According to the theoretical predictions, in the case of the smallest bombarding energy, the e^+e^- angular correlation could still be interpreted only by the internal pair creation process of the E1 multipolarity radiation since the excitation energy is very close to the threshold of X17 creation. In the case of higher energies it was already necessary to take into account the contribution of X17. In the case of different excitation energies, the location of the anomalous peak was constantly changing. At the same time the branching ratio compared to the E1 component of the radiation stayed constant. At the highest proton energy we were at off-resonance already, where we did not expect any anomaly.

With this, our experimental results confirmed the vectorial nature of the X17 particle and the protophobic interpretation of it, which could lead to the introduction of a fifth interaction and a better understanding of the properties of dark matter.

Our plans also include the study of the angular correlation of e^+e^- pairs generated by the decay of the giant dipole resonance (GDR) in the $^{11}B(p,e^+e^-)^{12}C$ nuclear reaction. In this case, the maximum of the anomaly related to the X17 is expected to be around 100°. Since GDR is observed in all nuclei, these results could trigger a systematic nuclear physics study of the X17 particle.

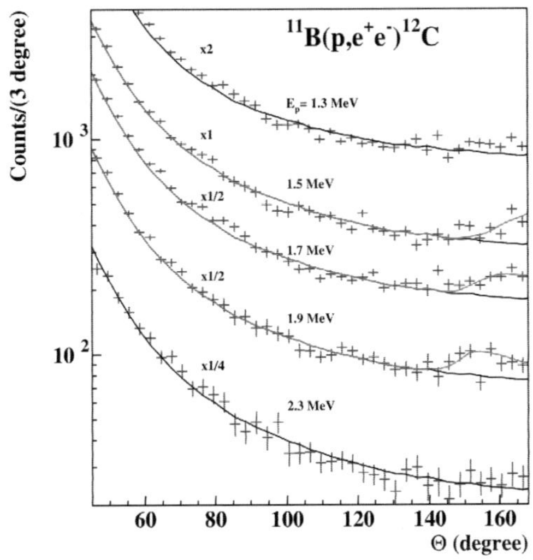

Figure 6. Angular correlations of the e^+e^- pairs obtained from the ^{11}B(p,γ)^{12}C reaction at different bombarding energies indicated in the figure. The simulated angular correlations for the X17 particle fitted to the data are also shown in red.

5. Acknowledgements

We wish to thank M. Viviani for providing us with their theoretical results. We wish to thank also Gy. Zilizi for providing us with the ^{3}H targets, Z. Pintye for the mechanical and J. Molnár for the electronic design of the experiment.

This work has been supported by the Hungarian NKFI Foundation No. K124810 and by the GINOP-2.3.3-15-2016-00034 and GINOP-2.3.3-15-2016-00034 grants.

References

[1] M. Pospelov, A. Ritz, and M.B. Voloshin, *Phys. Lett.* B **662** 53 (2008).
[2] T. Kitahara and Y. Yamamoto, *Phys. Rev.* D **95** 015008 (2017).
[3] A.J. Krasznahorkay et al., *Phys. Rev. Lett.* **116** 042501 (2016).
[4] M.E. Rose, Phys. Rev. **76**, (1949) 678. M.E. Rose, Phys. Rev. **78** 184 (1950).
[5] J. Feng et al., *Phys. Rev. Lett.* **117** 071803 (2016).
[6] J. Feng et al. *Phys. Rev.* D **95** 035017 (2017).
[7] Citations for Ref. [3]: https://inspirehep.net/search?ln=en&p=refersto%3Arecid%3A1358248&jrec=26&sf=earliestdate
[8] J. Gulyás et al. *Nucl. Instr. and Meth. in Phys. Res.* A **808**, 21 (2016).
[9] R. Brun, F. Bruyant, M. Maire, A.C. McPherson, P. Zanarini, Geant3 User's guide, Report number: CERN-DD-EE-84-01, (1987).
[10] K.E. Stiebing et al., J. Phys. G: Nucl. Part. Phys. **30** 165 (2004).
[11] U. Ellwanger and S. Moretti, J. High Energy Phys. **11** 39 (2016).
[12] Xilin Zhang, and Gerald A. Miller, Phys. Lett. **B 813** 136061 (2021).
[13] A. Krasznahorkay et al., Phys. Rev. C **104**, 044003 (2021).
[14] K.I. Hahn, C.R. Brune, and R.W. Kavanagh, Phys. Rev. C, 51 1624 (1995).
[15] D.R. Tilley, H.R. Weller, G.M. Hale, Nucl. Phys. **A541** 1 (1992).
[16] M. Viviani et al., Phys. Rev. C **105**, 014001 (2022).
[17] W. Verkerke and D. P. Kirkby, "The RooFit toolkit for data modeling," eConf C **0303241** (2003) MOLT007 [physics/0306116].

Journal of Physics: Conference Series **2391** (2022) 012008 doi:10.1088/1742-6596/2391/1/012008

Status of the X17 search in Montreal

G. Azuelos[1], D. Bryman[2], W.C. Chen[1], H. de Luz[3], L. Doria[4], A. Gupta[1], L-A. Hamel[1], M. Laurin[1], K. Leach[5], G. Lefebvre[1], J-P. Martin[1], A. Robinson[1], N. Starinski[1], R. Sykora[3], D. Tiwari[6], U. Wichoski[7], V. Zacek[1]

[1] Université de Montréal, Canada
[2] University of British Columbia, Canada
[3] Czech Technical University, Czech Republic
[4] University of Mainz, Germany
[5] Colorado School of Mines, USA
[6] University of Regina, Canada
[7] Laurentian University, Canada

E-mail: georges.azuelos@umontreal.ca

Abstract. At the Montreal Tandem accelerator, an experiment is being set up to measure internal pair creation from the decay of nuclear excited states using a multiwire proportional chamber and scintillator bars surrounding it from the Daphne experiment. The acceptance covers a solid angle of nearly 4π. Preamplifiers and the data acquisition hardware have been designed and tested. The ^{7}LiF target, mounted on an Al foil and water-cooled is in a thin carbon fiber section of the beamline. The experiment will focus at first on a measurement of the internal pair creation from the 18.15 MeV state of ^{8}Be with the aim of observing independently the X17 particle discovered by the ATOMKI experiment. Assuming the ATOMKI evaluation of the electron-pair production rate from X17, Geant4 simulation predicts observation of a clear signal after about 2 weeks of data taking with 2 μA proton beam. The IPC measurement could eventually be extended to the giant dipole resonance of ^{8}Be, as well as to other nuclei, in particular to ^{10}B.

1. Introduction

The Hungarian ATOMKI experiment [1, 2] has recently claimed observation, with a significance of more than 6 standard deviations, of a particle of mass around 17 MeV/c^2 in the spectrum of internal pair creation (IPC) from the decay of the 18.15 MeV state of ^{8}Be, as well as in the decay of states of ^{4}He near 20 MeV of excitation. This has aroused enormous interest and speculation on the nature of this so-called X17 particle. One description, consistent with present observations, would be a protophobic vector particle [3] for which the production rate depends principally, when off-resonance, on the E1 transition from direct proton capture [4, 5] .

This contribution gives an update on the preparations of an experiment, being set up in Montreal, for an independent measurement of IPC from ^{8}Be, confirming the existence of X17. The Université de Montréal 6 MV tandem Van de Graaff facility can deliver proton beams of 2μA, possibly up to 10 μA with an energy resolution of 2 keV for proton energies in the range 0.4-1.0 MeV. The 18.15 MeV 1^+ state of ^{8}Be is above the proton separation energy and decays principally by proton emission. Nevertheless, there is a small fraction of M1 electromagnetic decays to the 0^+ ground state. The IPC coefficient is 3.9 $\times 10^{-3}$ and the ratio of branching ratios $B(X \rightarrow e^+e^-)/B(^8Be^* \rightarrow \gamma)$ is 5.8×10^{-6}, according to the ATOMKI measurement. The production rate of the X17 particle is therefore extremely small. Nevertheless, it can be measured with a high significance since IPC is dominated by low invariant mass of the e^+e^- pair, and therefore a large boost and small opening angle, whereas the decay of the relatively heavy X17 particle gives a small boost and large opening angle to the e^+e^- pair. The observation of X17 requires, therefore, an experimental setup with large solid angle of acceptance, efficient rejection of gamma-ray background and precision measurement of the directions of the electron and positron.

2. Experimental setup

The experimental setup (Fig. 1) consists principally of a cylindrical multiwire proportional chamber (MWPC) of inner radius 6 cm and a length 36 cm, filled with 74% Ar and 26% CO_2 and surrounded by 16 scintillator bars of length 1.4 m arranged cylindrically at an inner radius of 16 cm. The chamber and the scintillator bars were previously used by the Daphne experiment at Saclay and Mainz. A number of phototubes are borrowed from TRIUMF. With its 192 anode wires with 2 mm spacing and 68 cathode strips of width 4 mm arranged at 45° to the anode wires, the angular resolution of the MWPC is around 2° FWHM assuming a 2mm diameter beam spot. It is made up of light material to reduce the effect of external pair creation (EPC) from gamma rays. Signals from the two ends of the scintillator bars allow a good measurement of the energy deposit of the electrons with an estimated resolution of $7\%/\sqrt{E(MeV)}$ as well as a position resolution of ~ 5 cm FWHM along the bar. The active detector volume covers a solid angle of 95% of 4π. A second, larger, MWPC is available, but will not be used in a first stage. A ^{7}LiF target will be evaporated on a thin Al foil which will be inclined at 45° to the beam line and held by a cooling rod. In the region of the target, the beamline is made of low density carbon fiber with a wall thickness of 0.8 mm. COMSOL simulations predict that the target temperature will be below 70°, assuming 20W heating from a proton beam of 10μA.

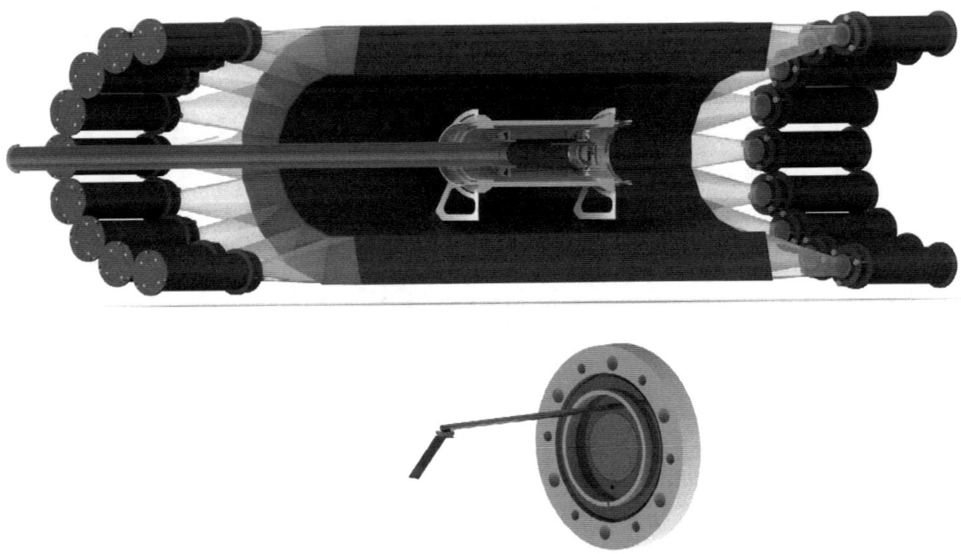

Figure 1. Basic experimental setup. Top: wire chamber surrounded by scintillator bars. Bottom: the target holder.

Preamplifiers for the MWPC wires and strips (Fig. 2) have been designed and tested. They will be placed close to the ends of the chamber. The signals are sent to VF48 ADC's and VT48 TDC's, designed in Montreal for an earlier experiment. The trigger (Fig. 2) is based on a coincidence of signals from at least 2 scintillator bars and two hits of the MWPC wires. Front end FPGA's perform baseline subtraction and gain correction from the photomultiplier tubes. The sum of energy depositions in the scintillator bars is calculated and a threshold is set for the trigger. The firmware for the data acquisition has been tested with a partial number of channels.

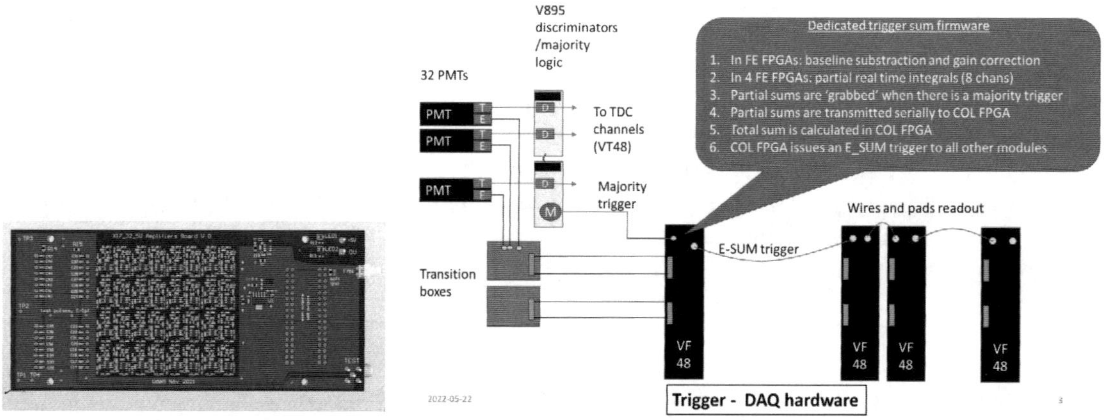

Figure 2. Charge preamplifier (left) and Level 1 trigger scheme (right).

3. Monte Carlo simulation

The Monte Carlo simulation with GEANT4 of the experimental setup is well advanced. The basic elements of the detector are included (see Fig. 3) and the resolution on the energy measurement in the scintillators and on the electron directions in the MWPC is taken into account. Thanks to the large solid angle acceptance of the experimental setup, the efficiency of reconstruction will have a smooth dependence on the electron pair opening angle. This would be an improvement over the ATOMKI setup, which consists of 6 isolated detectors [6]

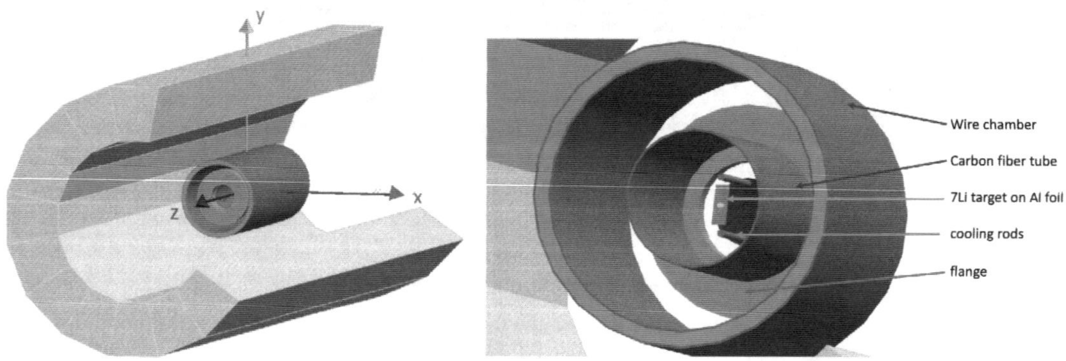

Figure 3. Geometry of the detector in Geant4 Monte Carlo.

The IPC spectrum from M1 as well as from E1 transitions are simulated, using expressions from Rose [7] and Stiebing [8]. The reconstruction of the X17 signal will adopt the strategy of the ATOMKI experiment. It is principally based on the measurement of the opening angle between the e^+ and the e^- since it is directly correlated with the invariant mass of the pair. For a low invariant mass, which is where most of the IPC spectrum lies, large opening angles are also correlated with a large asymmetry between the e^+ and the e^- energies. A selection cut on this asymmetry is therefore also applied: $y \equiv |E_1 - E_2|/(E_1 + E_2)$. Finally, to reduce background from poorly reconstructed electron pairs, an invariant mass selection will also be applied. A machine learning algorithm is being considered for selection of signal over background.

Based on the spectrum of gamma rays measured in a beam test with a BGO detector, and extrapolation to 95% of 4π acceptance, and based on the measured rate of the 478 keV photons from the $^7Li(p,p'\gamma)^7Li$ reaction, and assuming the IPC to gamma ratio above, it is estimated that the rate of X17 production will be $\sim 9\ \mathrm{h}^{-1}$ with a 2μA proton beam. Under these conditions, the IPC background, in the region of interest, corresponding to an opening angle of the e^+e^- pair of around 140°, is expected to be about 15 h^{-1}. The predicted opening angle distributions, based on the X/IPC ratio and assuming a pure M1 or a mixture of M1 and E1 transitions, are shown in Fig. 4. The background depends very sensitively on the ratio of E1/M1 [5] and will need to be fitted in a control region outside the expected excess. For these plots, the asymmetry parameter is limited to $y < 0.45$ and the invariant mass m_{ee} is greater than 12 MeV. The distributions do not account for statistical fluctuations. They are normalized to one X17 particle and assume good calibration of the energy depositions in the different scintillators. Clearly, with 1-2 weeks of data taking, with a X17 production of 9 h^{-1}, the signal should be very significant.

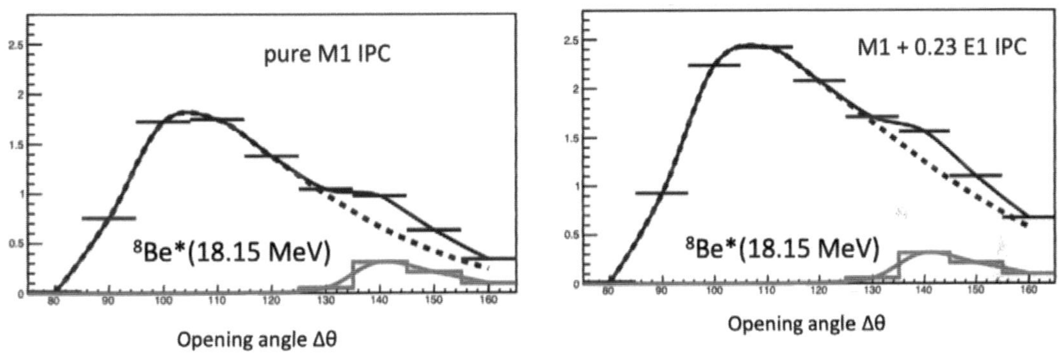

Figure 4. Predicted opening angle distributions (see text for details).

4. Conclusion and outlook

An experimental setup for an independent confirmation of the existence of the X17 particle is in preparation in Montreal. All the elements are being put in place: target, beamline, MWPC, scintillator bars. The electronic data acquisition and readout are nearly ready. Beam tests are expected to start in autumn 2022, followed by calibrations and tuning. Data taking should take place at the end of 2022. The large solid angle acceptance will allow a clear measurement of the opening angle of the electron pair. It will also help in defining the quantum numbers of the X17, as the distribution, for different forward angles of the electrons, depends on the nature of the particle [9].

As mentioned in the introduction, IPC in highly excited states of other nuclei can help to confirm the existence of X17 and understand its nature. One possibility being considered is the E1 transition from the 19.3 MeV state of ^{10}B, produced with the reaction $^7Li(^3He,\gamma)^{10}B$. Compared to the $^8Be^*$ transition, because of the higher energy, the comparable production cross section, and the E1 nature of the transition, the signal can be significantly more prominent. Also, since X production off resonance is dominated by direct proton capture, it is expected to be strongly enhanced at higher proton energies populating the giant dipole resonance of $^8Be^*$ [10].

References

[1] A. J. Krasznahorkay, M. Csatlós, L. Csige, Z. Gácsi, J. Gulyás, M. Hunyadi, T. J. Ketel, A. Krasznahorkay, I. Kuti and B. M. Nyakó, *et al.* Phys. Rev. Lett. **116** (2016) no.4, 042501 doi:10.1103/PhysRevLett.116.042501 [arXiv:1504.01527 [nucl-ex]].
[2] A. J. Krasznahorkay, M. Csatlós, L. Csige, J. Gulyás, M. Koszta, B. Szihalmi, J. Timár, D. S. Firak, Á. Nagy and N. J. Sas, *et al.* [arXiv:1910.10459 [nucl-ex]].
[3] J. L. Feng, B. Fornal, I. Galon, S. Gardner, J. Smolinsky, T. M. P. Tait and P. Tanedo, Phys. Rev. D **95** (2017) no.3, 035017 doi:10.1103/PhysRevD.95.035017 [arXiv:1608.03591 [hep-ph]].
[4] X. Zhang and G. A. Miller, Phys. Lett. B **813** (2021), 136061 doi:10.1016/j.physletb.2021.136061 [arXiv:2008.11288 [hep-ph]].
[5] N. J. Sas, A. J. Krasznahorkay, M. Csatlós, J. Gulyás, B. Kertész, A. Krasznahorkay, J. Molnár, I. Rajta, J. Timár and I. Vajda, *et al.* [arXiv:2205.07744 [nucl-ex]].
[6] J. Gulyás, T. J. Ketel, A. J. Krasznahorkay, M. Csatlós, L. Csige, Z. Gácsi, M. Hunyadi, A. Krasznahorkay, A. Vitéz and T. G. Tornyi, Nucl. Instrum. Meth. A **808** (2016), 21-28 doi:10.1016/j.nima.2015.11.009 [arXiv:1504.00489 [nucl-ex]].
[7] M. E. Rose, Phys. Rev. **76** (1949), 678-681 [erratum: Phys. Rev. **78** (1950), 184-184] doi:10.1103/PhysRev.78.184
[8] K. E. Stiebing, F. W. N. de Boer, O. Frohlich, H. Bokemeyer, K. A. Muller, K. Bethge and J. van Klinken, J. Phys. G **30** (2004), 165-180 doi:10.1088/0954-3899/30/2/014 [arXiv:nucl-ex/0311002 [nucl-ex]].
[9] M. Viviani, E. Filandri, L. Girlanda, C. Gustavino, A. Kievsky, L. E. Marcucci and R. Schiavilla, Phys. Rev. C **105** (2022) no.1, 014001 doi:10.1103/PhysRevC.105.014001 [arXiv:2104.07808 [nucl-th]].
[10] G. A. Fisher, P. Paul, F. Riess and S. S. Hanna, Phys. Rev. C **14** (1976), 28-36 doi:10.1103/PhysRevC.14.28

Theoretical study of the ^{3}H$(p, e^+e^-)^4$He and ^{3}He$(n, e^+e^-)^4$He processes and the X17 anomaly

M. Viviani
[1] INFN-Pisa
E-mail: michele.viviani@pi.infn.it

Abstract. The unexpected observation of eccess of events in the angular distribution of e^-e^+ pairs in various nuclear transitions has spurred a large interest, both experimentally and theoretically. This eccess has been interpreted as the possible existence of a new bosonic particle, the so-called X17, of mass around 17 MeV. In the present work, we investigate theoretically the possible effects of the presence of the X17 in the ^{3}H$(p, e^+e^-)^4$He and ^{3}He$(n, e^+e^-)^4$He reactions. For these processes it is possible to compute accurate *ab initio* bound- and continuum-states wave functions, so the existence of X17 can be unambigously revealed. Moreover, by exploiting the rich structure of the ^{4}He spectrum, it is possible to determine its quantum number, as, for example, if it is either a scalar, a pseudoscalar, a vector, or an axial particle.

1. Introduction

Recently there were claims [1, 2, 3, 4] that an unknown particle (denoted as "X17") had been observed in the processes ^{7}Li$(p, e^+e^-)^8$Be and ^{3}H$(p, e^+e^-)^4$He at the ATOMKI experimental facility situated in Debrecen (Hungary). These claims were based on a $\approx 7\sigma$ excess of events in the angular distribution of leptonic pairs produced in these reactions, which have a Q-value of about 20 MeV. The excess—referred to below as the "anomaly"—could be explained by positing the emission of an unknown boson with a mass of about 17 MeV decaying into e^+e^- pairs.

The possible existence of a new kind of low mass particle (at the MeV scale) is a problem of current and intense theoretical and experimental interest (see, for example, Ref. [5] and references therein). This interest is, in fact, part of a broader effort aimed at identifying dark matter (DM). The search for bosonic DM candidates had already started by several years, by attempting to establish the possible existence of additional forces (beyond gravity), mediated by these bosons [6] between DM and visible matter. To one such class of particles belongs the so-called "dark-photon", namely a boson of mass M_X having the same quantum numbers as the photon, and interacting with a Standard Model (SM) fermion f with a coupling constant given by $\varepsilon\, q_f$, where q_f is the fermion electric charge. Following several years of experimental searches, "exclusion plots" in the ε-M_X parameter space were produced, restricting more and more the allowed region [5, 7]. One of the most stringent limits was provided by the NA48/2 experiment [8].

The observation of the ^{8}Be and ^{4}He anomalies by the ATOMKI group soon spurred several theoretical studies. In Ref. [9], the possibility that the X17 could be a vector boson was investigated in detail. In order to circumvent the NA48/2 limit, it was conjectured that the X17 could be "proto-phobic", namely that it would couple much more weakly to the proton than to the neutron [9]. Many other theoretical studies were published aftewards [10, 11, 12, 13, 14, 15, 16, 17, 18, 19, 20, 21, 22, 23, 24].

On the experimental side, there are several experiments (MEGII [25], DarkLight [26], SHiP [27], the "Montreal X-17 Project" [28], TREK/E36 [29] and others [5]) planning specifically to search for such a light boson. In addition, large collaborations, such as BelleII [30], NA64 [31], and others, are dedicating part of their efforts in an attempt to clarify this issue. Recently, the reaction $^3\text{He}(n,e^+e^-)^4\text{He}$ has been proposed as a possible new process where the presence of X17 can be observed and, eventually, its properties can be determined [32].

In the present contribution, we present a study of the $^3\text{H}(p,e^+e^-)^4\text{He}$ and $^3\text{He}(n,e^+e^-)^4\text{He}$ reactions. The advantage of studying these processes is that for them it is possible to perform accurate *ab initio* calculations to describe bound- and continuum-states. We use the hyperspherical-harmonics (HH) method to determine them [33, 34, 35], by taking fully into account of the four-body dynamics. The employed nuclear Hamiltonians are very accurate. They consist of non-relativistic kinetic energy, two-nucleon ($2N$) and three-nucleon ($3N$) interactions derived within the framework of chiral effective field theory (χEFT) at fourth order [36, 37, 38, 39, 40]. To perform this study, we need of accurate χEFT electromagnetic currents, as well. We have taken them from Refs. [41, 42, 43, 44, 45].

Regarding the X17 interaction, we consider the general case of a Yukawa-like interaction between this boson and a SM fermion of species f (specifically, u and d quarks, and electrons) with the coupling constant expressed as $\varepsilon_f\, e$, where e (>0) is the unit electric charge. The X17 boson must decay promptly in e^+-e^- pairs for these to be detected inside the experimental setup. This observation actually introduces a *lower limit* to the possible values of ε_e. These limits are also established by various electron beam-dump experiments (see, for example, Ref. [46] and references therein). For $M_X \approx 17$ MeV, the most stringent lower bound, $|\varepsilon_e| > 2\times 10^{-4}$, comes from the SLAC E141 experiment [47], while the upper bound $|\varepsilon_e| < 2\times 10^{-3}$ has been set by the KLOE-2 experiment [48]. However, we will show that in the considered nuclear processes there is no sensitivity to ε_e. In the following, we will assume $\varepsilon_e = 10^{-3}$ and study how the angular distribution of the two leptons is affected by the values of the spin and parity of the X17 boson.

The work described in this contribution is based on the study presented in Ref. [49]. In Sect. 2, a brief description of the theoretical formalism is given. Then, in Sect. 3 the results of the calculations are reported and discussed. Finally, in the last section, the perspectives of this study are given.

2. Theoretical analysis

In this work, the X17 interaction Lagrangian density at the energy scale $\Lambda_H \sim 1$ GeV is considered to be

$$\mathcal{L}^c_X(x) = e \sum_{f=e,u,d,\ldots} \varepsilon_f\, \overline{f}(x)\, \Gamma^c\, f(x)\, X_c(x)\ , \tag{1}$$

where $c = S, P, V$, or A for a scalar, pseusoscalar, vector, or axial boson, respectively. Correspondingly,

$$\Gamma^{c=S,P,V,A} = 1, i\,\gamma^5, \gamma^\mu, \gamma^\mu\,\gamma^5\ . \tag{2}$$

In eq. (1), the sum runs over the lightest fermions of the SM[1] and $X_c(x) = X(x)$ for $c = S, P$ and $X_c(x) = X_\mu(x)$ for $c = V, A$ represents the X17 field. The various coupling constant ε_f are written in units of the electric charge $e > 0$ ($e^2 = 4\pi\alpha$).

[1] Here we consider the most general and simple "effective" Lagrangian between X17 and the lightest particles of the SM. All the other degrees of freedom (heavy quarks, the Higgs, Z and W bosons, etc.) are considered to be "integrated out". In principle, it is possible, starting from a BSM theory at energy scale $\gg \Lambda_H$, to evolve the various interactions terms to the 1 GeV scale, resulting in a Lagrangian as given in Eq. (1). Once the various coefficients of this Lagrangian ε_e, ε_u, ε_d, ... be eventually determined, one can work out back to the high-energy scale in order to check each particular BSM theory.

Journal of Physics: Conference Series **2391** (2022) 012009 doi:10.1088/1742-6596/2391/1/012009

Starting from this Lagrangian, it is possible to derive nucleon-X17 interaction Lagrangian densities in the framework of χEFT. By retaining only leading-order contributions (and selected subleading ones in the vector and pseudoscalar cases), one obtains (for a detailed derivation, see Ref. [49])

$$\begin{aligned} \mathcal{L}_X^S(x) &= e\,\overline{N}(x)[\eta_0^S + \eta_z^S\,\tau_3]N(x)\,X(x)\,, & (3)\\ \mathcal{L}_X^P(x) &= e\,\eta_z^P\,\pi_3(x)\,X(x) + e\,\eta_0^P\overline{N}(x)i\,\gamma^5 N(x)\,X(x)\,, & (4)\\ \mathcal{L}_X^V(x) &= e\,\overline{N}(x)[\eta_0^V + \eta_z^V\,\tau_3]\gamma^\mu\,N(x)\,X_\mu(x) & (5)\\ &\quad + \frac{e}{4\,m_N}\,\overline{N}(x)[\kappa_0\eta_0^V + \kappa_z\eta_z^V\,\tau_3]\sigma^{\mu\nu}\,N(x)F_{\mu\nu}^X(x)\,, \\ \mathcal{L}_X^A(x) &= e\,\overline{N}(x)[\eta_0^A + \eta_z^A\,\tau_3]\gamma^\mu\gamma^5\,N(x)\,X_\mu(x)\,, & (6) \end{aligned}$$

where m_N is the nucleon mass, $N(x)$ is the iso-doublet of nucleon fields, $\pi_3(x)$ is the third component of the triplet of pion fields, and $F_{\mu\nu}^X(x) = \partial_\mu\,X_\nu(x) - \partial_\nu\,X_\mu(x)$ is the X17 field tensor. The hadron-X17 coupling constants η_0^c and η_z^c are linear combinations of the quark-X17 coupling constants ε_u and ε_d, see Ref. [49] for their explicit expressions. In the vector case, we have included also the subleading term proportional to $F_{\mu\nu}^X$ and where

$$\kappa_0 = \kappa_p + \kappa_n\,, \qquad \kappa_z = \kappa_p - \kappa_n\,, \tag{7}$$

κ_p and κ_n being the anomalous magnetic moments of the proton and neutron, respectively. In the pseudoscalar case, the interaction at leading order in the power counting originates from the direct coupling of the X17 to the pion. However, since the associated coupling constant is expected to be suppressed [17, 18], we have also considered an isoscalar coupling of the X17 to the nucleon, even though it is subleading, at least nominally, in the χEFT power counting relative to the isovector one. As per the axial case, the leading order $\mathcal{L}_X^A(x)$ contains an additional term of the form $\partial^\mu\pi_3(x)\,X_\mu(x)$, which we have dropped. In fact, this term leads to a X17-nucleon current proportional to q^μ/m_π^2 (for low momentum transfers) which, when contracted with the lepton axial current, produces a contribution proportional to $(m_e/m_\pi)^2$, and hence negligible when compared to that resulting from the X17 direct coupling to the nucleon.

Starting from the expressions above, at first order in perturbation theory, the amplitude for the emission of an e^+-e^- pair between nuclear initial and final states is written in general as

$$T_{fi} = T_{fi}^{EM} + T_{fi}^{cX}\,, \tag{8}$$

where T_{fi}^{EM} is the one-photon-exchange amplitude written as

$$T_{fi}^{EM} = 4\pi\alpha\,\frac{(\overline{u}_-\,\gamma_\mu\,v_+)\,j_{fi}^\mu}{q^\mu\,q_\mu}\,. \tag{9}$$

Above, α is the fine structure constant, q^μ is the four-momentum transfer defined as the sum of the outgoing-lepton four momenta, u_- and v_+ are, respectively, the electron and positron spinors, and j_{fi}^μ is the matrix element of the nuclear electromagnetic current between the initial and final nuclear states (here, either the p+^{3}H or n+^{3}He scattering state and ^{4}He ground-state).

The term T_{fi}^{cX} is the amplitude with the exchange of a unknown boson X17 of type $c = S, P, V, A$, namely

$$T_{fi}^{cX} = 4\pi\alpha\,\frac{\varepsilon_e\,(\overline{u}_-\,\Gamma_c\,v_+)\,j_{fi}^{cX}}{q^\mu\,q_\mu - M_X^2}\,, \tag{10}$$

where M_X is the mass of the X17 particle, and j_{fi}^{cX} represents the matrix element of the X17-induced nuclear current, which implicitly depends on the X17-hadron coupling constants η_0^c and η_z^c.

In the laboratory frame, the initial state consists of an incoming proton or neutron of momentum $\boldsymbol{p}$ and spin projection m_1, and a bound ^{3}H or ^{3}He cluster in spin state m_3 at rest. Its wave function $\Psi^{(\gamma)}_{m_3,m_1}(\boldsymbol{p})$ is such that, in the asymptotic region of large separation $\boldsymbol{y}_\ell$ between the isolated nucleon (particle ℓ) and the trinucleon cluster (particles ijk), it reduces to

$$\Psi_{m_3,m_1}(\boldsymbol{p}) \longrightarrow \frac{1}{\sqrt{4}} \sum_{\ell=1}^{4} \phi^{m_3}(ijk)\chi^{m_1}(\ell)\, \Phi^{(\gamma)}_{\boldsymbol{p}}(\boldsymbol{y}_\ell) \; , \tag{11}$$

where $\Phi_{\boldsymbol{p}}(\boldsymbol{y}_\ell)$ is either a Coulomb distorted wave or simply the plane wave $e^{i\boldsymbol{p}\cdot\boldsymbol{y}_\ell}$ depending on whether we are dealing with the p+^{3}H or n+^{3}He state (correspondingly, ϕ^{m_3} is the ^{3}H or ^{3}He bound state wave function). The final state consists of the lepton pair—with the e^- having momentum (energy) $\boldsymbol{k}$ (ϵ) and spin s, and the e^+ having momentum (energy) $\boldsymbol{k}'$ (ϵ') and spin s'—and the ^{4}He ground state recoiling with momentum $\boldsymbol{p}-\boldsymbol{k}-\boldsymbol{k}'$. Energy conservation requires

$$\epsilon + \epsilon' + \frac{(\boldsymbol{p}-\boldsymbol{k}-\boldsymbol{k}')^2}{2\,M} = E_0 \; , \tag{12}$$

where M is the rest mass of the ^{4}He ground state. Further, we have defined $E_0 = T_p + B_4 - B_3 \approx 20$ MeV, where T_p is the kinetic energy of the incident nucleon, and B_3 and B_4 are the binding energies of, respectively, the bound three-nucleon cluster and ^{4}He.

The general expression of the five-fold differential cross section can be schematically written as

$$\begin{aligned} \frac{d^5\sigma}{d\epsilon\, d\hat{\boldsymbol{k}}\, d\hat{\boldsymbol{k}}'} &= \sigma_{EM}(\epsilon,\hat{\boldsymbol{k}},\hat{\boldsymbol{k}}') + \frac{\varepsilon_e R_X(\epsilon,\hat{\boldsymbol{k}},\hat{\boldsymbol{k}}')}{D_X} + c.c. + \frac{\varepsilon_e^2 R_{XX}(\epsilon,\hat{\boldsymbol{k}},\hat{\boldsymbol{k}}')}{|D_X|^2} \; , \\ &= \sigma_{EM}(\epsilon,\hat{\boldsymbol{k}},\hat{\boldsymbol{k}}') + \frac{\varepsilon_e R_X(\epsilon,\hat{\boldsymbol{k}},\hat{\boldsymbol{k}}')D_X^* + c.c. + \varepsilon_e^2 R_{XX}(\epsilon,\hat{\boldsymbol{k}},\hat{\boldsymbol{k}}')}{|D_X|^2} \; , \end{aligned} \tag{13}$$

where $D_X = Q^2 - M_X^2$ (with $Q^2 = q^\mu q_\mu$) and the three terms denote the contributions coming solely from electromagnetic currents, the interference between electromagnetic and X17-induced currents, and purely X17-induced currents, respectively. The positron energy ϵ' is fixed by energy conservation. In the laboratory coordinate system, where the z-axis is oriented along the incident beam momentum $\boldsymbol{p}$, the spherical angles specifying the $\hat{\boldsymbol{k}}$ ($\hat{\boldsymbol{k}}'$) direction are denoted as θ and ϕ (θ' and ϕ'), and

$$\hat{\boldsymbol{k}}\cdot\hat{\boldsymbol{k}}' \equiv \cos\theta_{ee} = \cos\theta\cos\theta' + \sin\theta\sin\theta'\cos(\phi'-\phi) \; . \tag{14}$$

In Eq. (13) we have made explicit the dependence on the X17-electron coupling constant ε_e. The quantity R_X (R_{XX}) depends linearly (quadratically) on the X17-hadron coupling constants $\eta^c_{0,z}$.

The four-fold differential cross section is obtained by integrating over the electron energy,

$$\frac{d^4\sigma}{d\hat{\boldsymbol{k}}\, d\hat{\boldsymbol{k}}'} = \int_{m_e}^{\epsilon_{max}} d\epsilon \, \frac{d^5\sigma}{d\epsilon\, d\hat{\boldsymbol{k}}\, d\hat{\boldsymbol{k}}'} \; , \tag{15}$$

where the maximum allowed energy ϵ_{max} is obtained from the solution of Eq. (12) for the case of a vanishing positron momentum; in fact, since the kinetic energy of the recoiling ^{4}He is tiny, ϵ_{max} is close to $E_0 - m_e$.

To account for the decay of X17, we introduce a width Γ_X and will make in Eq. (13) the replacement

$$M_X \longrightarrow M_X - i\,\Gamma_X/2 \; . \tag{16}$$

Hence, $D_X = Q^2 - M_X^2 + \frac{\Gamma_X^2}{4} + iM_X\Gamma_X$. Assuming that the predominant decay channel of X17 is in e^+e^- [9], then $\Gamma_X \sim \alpha\,\varepsilon_e^2\,M_X \ll M_X$, given the current bounds on ε_e ($\sim 10^{-3}$ as discussed in the Introduction). Therefore $D_X \approx Q^2 - M_X^2 + iM_X\Gamma_X$. Now, the X17 contribution to the cross section (the term $\sim 1/|D_X|^2$) is only sizeable where $Q^2 - M_X^2 = 0$ [49]. We can distinguish two kinematical regions of ϵ, $\hat{\boldsymbol{k}}$, and $\hat{\boldsymbol{k}}'$ values:

- Region A, where the $Q^2 - M_X^2 = 0$ is never satisfied (for example, for $\theta = \theta' = 90$ deg and $\theta_{ee} < 112$ deg); in that region, the contribution of the X17 is always negligible with respect to σ_{EM}, therefore here we can safely assume that

$$\frac{d^5\sigma}{d\epsilon\, d\hat{\boldsymbol{k}}\, d\hat{\boldsymbol{k}}'} = \sigma_{EM}(\epsilon, \hat{\boldsymbol{k}}, \hat{\boldsymbol{k}}') \quad \text{region A} . \tag{17}$$

- Region B, where $Q^2 - M_X^2 = 0$ is satisfied for some values of ϵ, $\hat{\boldsymbol{k}}$, and $\hat{\boldsymbol{k}}'$. For fixed values of $\hat{\boldsymbol{k}}$, and $\hat{\boldsymbol{k}}'$, the condition $Q^2 - M_X^2 = 0$ is verified for two values of ϵ, denoted in the following as ϵ_i, $i = 1, 2$ [49]; since Γ_X is very small, $1/|D_X|^2$ is always negligible except for $\epsilon \approx \epsilon_i$, where it assumes the form of a very narrow Lorentzian, which can be very well represented by a delta function; namely,

$$\frac{1}{|D_X|^2} \longrightarrow \sum_{i=1,2} \frac{\gamma_i}{\varepsilon_e^2}\delta(\epsilon - \epsilon_i) , \tag{18}$$

where γ_i are factors independent on ϵ and ε_e [49] (clearly, ϵ_i and γ_i depend on the given choice of $\hat{\boldsymbol{k}}, \hat{\boldsymbol{k}}'$). It is worthwhile to point out that, in region B, the interference contribution (proportional to R_X) is always negligible relative to the R_{XX} term [49]. Therefore, we can write

$$\frac{d^5\sigma}{d\epsilon\, d\hat{\boldsymbol{k}}\, d\hat{\boldsymbol{k}}'} = \sigma_{EM}(\epsilon, \hat{\boldsymbol{k}}, \hat{\boldsymbol{k}}') + \sum_{i=1,2} R_{XX}(\epsilon_i, \hat{\boldsymbol{k}}, \hat{\boldsymbol{k}}')\gamma_i\delta(\epsilon - \epsilon_i) , \quad \text{region B} . \tag{19}$$

Note that, in the present tree-level treatment of the X17 width, the cross-section becomes independent on ε_e.

3. Numerical results

In this contribution, we compare our calculation with the ATOMKI data for the $^3\text{H}(p, e^-e^+)^4\text{He}$ reaction reported in Refs. [2] and [3]. We consider the four-fold differential cross sections obtained by integrating over the electron energy, with the remaining kinematical variables in the same configuration of the experiment (that is, the momenta of the lepton pair in the plane orthogonal to the incident proton momentum).

To take into account the finite experimental angular resolution, we have folded the calculated four-folded differential cross section with a Gaussian. Explictly, defined as $\sigma(\theta_{ee})$ the four-fold differential cross section calculated for the two leptons emitted in such a perpendicular plane, θ_{ee} being the angle between $\boldsymbol{k}$ and $\boldsymbol{k}'$, we compute

$$\overline{\sigma}(\theta_{ee}) = \frac{1}{\sqrt{2\pi\Delta^2}} \int d\theta'_{ee}\, \sigma(\theta'_{ee}) \exp\left(-\frac{(\theta'_{ee} - \theta_{ee})^2}{2\Delta^2}\right) , \tag{20}$$

where Δ represents the finite angular resolution. In this work, we have adopted the value $\Delta = 5$ deg (for more details, see Ref. [49]). Moreover, since the ATOMKI cross section measurements are unnormalized, we rescale them to match the calculated values for $\theta_{ee} < 90°$, where the cross section is dominated by the purely electromagnetic amplitude (we are well inside the kinematical region A defined above).

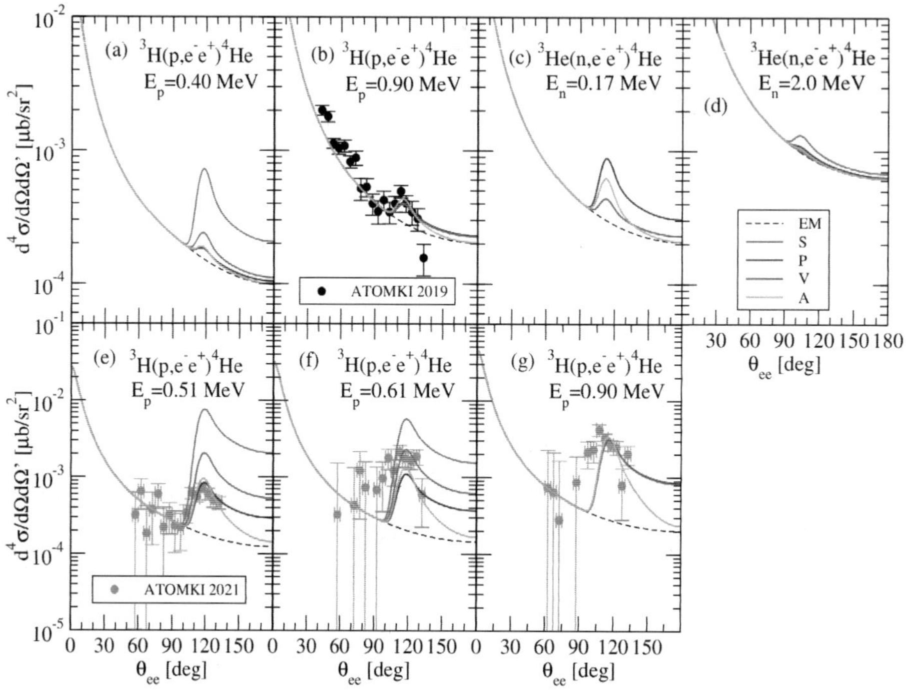

Figure 1. The averaged four-fold differential cross section $\overline{\sigma}(\theta_{ee})$ for the ^{3}H$(p,e^-e^+)^4$He and ^{3}He$(n,e^-e^+)^4$He processes at different incident nucleon energies for the configuration in which the e^+ and e^- momenta are in the plane orthogonal to the incident nucleon momentum and as function of the angle θ_{ee} between them. In all cases, we have taken $M_X = 17$ MeV and Γ_X as given from the X17 decay in e^-e^+. Panels (a), (b), (e), (f), and (g) report the results obtained for the ^{3}H$(p,e^-e^+)^4$He reaction, while panels (c) and (d) the results obtained for the ^{3}He$(n,e^-e^+)^4$He reaction. Moreover, the curves labeled S, P, V, and A show the results obtained by including the exchange of a scalar, pseudoscalar, vector, and axial X17, respectively. The dashed (black) curves show the results obtained by including the electromagnetic amplitude only. The results shown in the upper panels have been obtained by adjusting the coupling constants in order to reproduce the 2019 ATOMKI ^{3}H$(p,e^-e^+)^4$He cross section data of Ref. [2] at the incident proton energy of 0.90 MeV, see panel (b), and then making predictions for the other cases, see panels (a), (c) and (d). The results shown in the lower panels have been obtained by adjusting the coupling constants in order to reproduce the 2021 ATOMKI ^{3}H$(p,e^-e^+)^4$He cross section data of Ref. [3] at the incident proton energy of 0.90 MeV, see panel (g), and then making predictions for the other cases, see panels (e) and (f). The calculations are based on the N3LO nucleon-nucleon interaction of Refs. [36, 37], augmented by the N2LO three-nucleon interaction defined in Ref. [38], and accompanying electromagnetic currents. [41, 42, 43].

First, we consider the 2019 ATOMKI data [2] taken at $E_p = 0.90$ MeV. For each of the assumed couplings, we constrain one of the coupling constants $\eta^c_{0,z}$ (or a combination of them) by fitting these ATOMKI data in the range $\theta_{ee} > 110°$, where the (purported) X17 signal has been observed (we take as before $M_X = 17$ MeV). For these angles we are inside the kinematical region B and therefore the cross section is given by Eq. (19). Using these values for $\eta^c_{0,z}$ coupling constants, we then compute the four-fold cross section $\overline{\sigma}(\theta_{ee})$ for the ^{3}H$(p,e^-e^+)^4$He and ^{3}He$(n,e^-e^+)^4$He processes at different incident nucleon energies.

The results obtained are reported in the upper panels of Fig. 1. In panels (a) and (b) of such a figure, the incident proton energies of 0.40 and 0.90 MeV correspond to energies E_0 in Eq. (12) of 20.12 and 20.50 MeV, respectively. Referring to Fig. 2, we see that the lower 20.12 MeV corresponds to the energy of the first 0^+ excited state, while the higher 20.50 MeV (the energy selected in the ATOMKI experiment) is just below the $n + {}^3$He threshold. The incident neutron energy of 0.17 MeV, see panel (c), corresponds to $E_0 = 20.69$ MeV, close to the energy of the 0^- excited state, while the incident neutron energy of 2.0 MeV, see panel (d), is on top of the 2^- excited state.

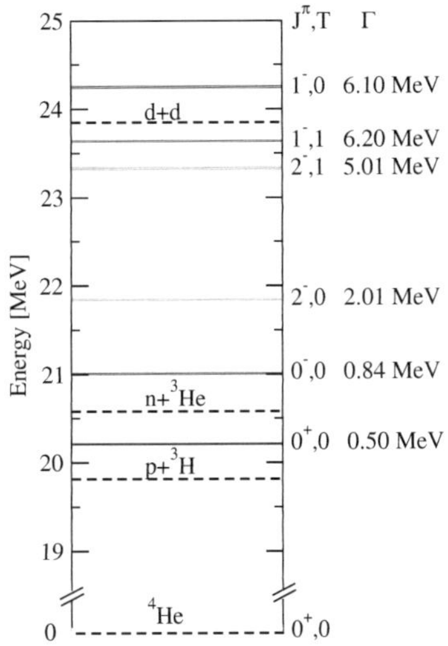

Figure 2. The low-energy spectrum of ^{4}He [50]. The dashed lines indicate the thresholds for the opening of the 1+3 and 2+2 channels, while the solid lines indicate the energies of the various resonances with corresponding widths, J^π, and isospin assignments to the right.

This structure of the ^{4}He low-energy spectrum and the selectivity of the X17-induced transition operator are reflected in the results of Fig. 1. In the following, we take into account that the isospin dependence of these operators is generally of the form $\eta_0^c + \eta_z^c\, \tau_{i,z}$. Let us consider the various cases one-by-one.

- S case. In this case we have selected $\eta_z^S = 0$ and fitted η_0^S to the ATOMKI data. In fact, the isoscalar component of the corresponding transition operator connects the (predominantly isoscalar) $J^\pi = 0^+$ resonance to the ^{4}He ground state. This transition is responsible for the prominent low-energy peak seen in Fig. 1; this peak rapidly fades away as the energy of the incident proton increases in the ^{3}H$(p, e^-e^+)^4$He process. It is barely visible in the ^{3}He$(n, e^-e^+)^4$He process, since the $n + {}^3$He threshold is already relatively far away from the 0^+ resonance energy, see Fig. 2. The isovector component of the S operator is suppressed and the corresponding cross section would not show any particular energy dependence.
- P case. Again, the purely isovector (pion-mediated) P transition operator has a small matrix element between the (predominantly $T = 0$) 0^- resonance and ^{4}He (0^+) ground state, we have ignored it altogether in the panels of Fig. 1 by setting $\eta_z^P = 0$—that is, by assuming a "piophobic" X17—and have instead considered only the contribution from the isoscalar P transition operator, which has a large matrix element between these 0^- and 0^+ states. As a consequence, a pronounced peak structure is seen when the energy approaches that of the 0^- resonance. As this energy increases well beyond the $n + {}^3$He threshold, additional significant contributions also come from the (relatively broad) 2^- resonances.

- V case. Here we have considered the case of a proto-phobic X17, namely we have assumed $\eta_0^V + \eta_z^V = 0$. The time component of the V operator has the same structure as the S operator. Its space component, however, produces large electric dipole (E_1) matrix elements between the 1^- scattering state and ^{4}He ground state (note in Fig. 2 the two wide 1^- resonances located at energies close to the $d+d$ threshold). These E_1 matrix elements are found to be much larger than the matrix elements due to the $0^+ \longrightarrow 0^+$ transition, and are the same for both photon and X17-induced amplitudes (modulo coupling constants, of course), see also Ref. [51] for a similar finding in connection with the ^{8}Be experiment. As these E_1 matrix elements are almost energy independent, once the value of η_0^V (clearly $\eta_z^V = -\eta_0^V$) has been constrained to reproduce the ATOMKI data at $E_p = 0.9$ MeV, the height of the peak in the full cross section is nearly constant relative to that in the purely electromagnetic cross section for all energies considered in this work. By contrast, the 1^+ scattering-state contributions are suppressed since they have magnetic-dipole (M_1) character. Moreover and most importantly, in this $^3S_1$ wave the Pauli principle prevents the four nucleons from coming close together. Even though there are well defined and fairly narrow 2^- resonances, the associated transitions, having M_2 character, are suppressed.
- A case. In this case we have retained only the isoscalar component of the transition operator, i.e. we have assumed $\eta_z^A = 0$. In fact, the isoscalar time component of the A operator induces big matrix elements between the (predominantly $T=0$) 0^- scattering state and ^{4}He ground state. This transition yields the dominant contribution, since now the $1^- \longrightarrow 0^+$ transition is M_1 and therefore suppressed by q relative to the $0^- \longrightarrow 0^+$ one above. The $1^+ \longrightarrow 0^+$ transition is E_1, but again inhibited by the Pauli repulsion. As a consequence, in the axial case the X17 peak is more pronounced for energies at which the 0^- state is populated, as for the P case. However, it should be noted that for large angles θ_{ee} (close to back-to-back configurations), the cross-section enhancement from A couplings is much reduced, due to the vanishing of the X17 $\rightarrow e^-e^+$ amplitude.

From this analysis, it is shown that the study of the ^{3}H$(p, e^-e^+)^4$He and ^{3}He$(n, e^-e^+)^4$He processes at various energy of the proton or neutron beam can be sufficient to disentangle the question of the spin and parity of X17. More information can be obtained by looking also at other kinematical configurations, in particular by detecting the leptons out of the perpendicular plane [49].

The extracted values of the coupling constants depend clearly on the parameters chosen to perform the angular smearing of the theoretical cross sections, and on the factor used to rescale the ATOMKI data. They can be found in Ref. [49].

Finally, we repeat the same procedure by comparing with the 2021 ATOMKI data of Ref. [3] (lower panels of Fig. 1). In this case, "background-free" ^{3}H$(p, e^-e^+)^4$He reaction data are also reported, obtained by subtracting the counting rate due to "external" pair conversion (EPC) processes. This EPC rate is estimated on the basis of a GEANT simulation of the processes where real photons emitted in the ^{3}H$(p, \gamma)^4$He radiative capture convert in lepton pairs by interacting with the experimental apparatus [3].

However, these data at angles $< 90°$ have large errors, making the matching between theory and experiment in this angular range rather problematic. Anyway, once the 2021 ATOMKI data have been rescaled, we again constrain one of the coupling constants $\eta_{0,z}^c$ (or a combination of them) by fitting the reported X17 peak at $E_p = 0.90$ Mev, see panel (g), and then compare with the data obtained at 0.51 and 0.61 MeV. As it can be seen by inspecting panels (e) and (f) of Fig. 1, the data at $E_p = 0.51$ and 0.61 MeV seem to be better reproduced when the X17 is either a pseudoscalar or an axial boson. As the energy decreases, the prominence of the peak relative to the baseline of e^+-e^- pairs produced by purely EM transitions is reduced, suggesting that the relevant matrix element may be that connecting the 0^- resonance to the ^{4}He ground state. However, we remark again that the large errors in the experimental data for $\theta_{ee} < 90$ deg render

Journal of Physics: Conference Series **2391** (2022) 012009 doi:10.1088/1742-6596/2391/1/012009

problematic the matching of these data to the photon-only cross sections. So these conclusions have to be considered as provisional.

4. Concluding remarks

A major objective of the present work has been to provide an accurate treatment of the e^+-e^- pair production process in the four-nucleon system, based on the one-photon-exchange approximation and a state-of-the-art χEFT description of nuclear interactions and electromagnetic currents. The initial $3+1$ scattering-state and ^{4}He bound-state wave functions have been obtained from HH solutions of the Schrödinger equation. In particular, these solutions fully account for the coupling among different (energetically open) channels and for the presence of resonances observed in the $A=4$ low-energy spectrum.

In the kinematics of the ATOMKI experiment where the e^+-e^- pair is detected in the plane orthogonal to the incident nucleon momentum, the Fig. 1 suggests that a systematic study of the cross section as function of both the opening angle and electron energy might allow us to discriminate among the different hypotheses regarding the nature of X17. Such a study is planned for the ^{3}He$(n, e^+e^-)^4$He process, currently in the feasibility and development phase at the CERN n_ToF facility [32].

Lastly, in order to further investigate the robustness of the currently predicted electromagnetic cross sections, we also plan to go beyond the one-photon-exchange approximation and include higher-order QED corrections in our treatment of pair production in the $A=4$ system. In fact, in Refs. [52, 53], the possibility that the peak seen in the ^{8}Be experiment could be caused by higher-order QED effects, beyond the one-photon-exchange approximation, has been investigated. It is not clear if such an explanation will remain valid in the case of the ^{4}He experiment, where a sharper peak is observed at a considerably smaller opening angle, around 112°.

References

[1] Krasznahorkay A J *et al.* 2016 *Phys. Rev. Lett.* **116** 042501 (*Preprint* 1504.01527)
[2] Krasznahorkay A *et al.* 2019 (*Preprint* 1910.10459)
[3] Krasznahorkay A J, Csatlós M, Csige L, Gulyás J, Krasznahorkay A, Nyakó B M, Rajta I, Timár J, Vajda I and Sas N J 2021 *Phys. Rev. C* **104** 044003 (*Preprint* 2104.10075)
[4] Sas N J *et al.* 2022 (*Preprint* 2205.07744)
[5] Battaglieri M *et al.* 2017 US Cosmic Visions: New Ideas in Dark Matter 2017: Community Report *U.S . Cosmic Visions: New Ideas in Dark Matter* (*Preprint* 1707.04591)
[6] Pospelov M, Ritz A and Voloshin M 2008 *Physics Letters B* **662** 53 – 61 ISSN 0370-2693 URL http://www.sciencedirect.com/science/article/pii/S0370269308002402
[7] Banerjee D *et al.* (NA64) 2018 *Phys. Rev. Lett.* **120** 231802 (*Preprint* 1803.07748)
[8] Batley J *et al.* (NA48/2) 2015 *Phys. Lett. B* **746** 178–185 (*Preprint* 1504.00607)
[9] Feng J L, Fornal B, Galon I, Gardner S, Smolinsky J, Tait T M P and Tanedo P 2016 *Phys. Rev. Lett.* **117** 071803 (*Preprint* 1604.07411)
[10] Zhang X and Miller G A 2017 *Phys. Lett. B* **773** 159–165 (*Preprint* 1703.04588)
[11] Ellwanger U and Moretti S 2016 *J. High Energ. Phys.* **2016** 39 (*Preprint* 1609.01669)
[12] Fornal B 2017 *Int. J. Mod. Phys. A* **32** 1730020 (*Preprint* 1707.09749)
[13] Dror J A, Lasenby R and Pospelov M 2017 *Phys. Rev. Lett.* **119** 141803 (*Preprint* 1705.06726)
[14] Delle Rose L, Khalil S and Moretti S 2017 *Phys. Rev. D* **96** 115024 (*Preprint* 1704.03436)
[15] Delle Rose L, Khalil S, King S J, Moretti S and Thabt A M 2019 *Phys. Rev. D* **99** 055022 (*Preprint* 1811.07953)
[16] Delle Rose L, Khalil S, King S J and Moretti S 2019 *Front. in Phys.* **7** 73 (*Preprint* 1812.05497)
[17] Alves D S M and Weiner N 2018 *JHEP* **07** 092 (*Preprint* 1710.03764)
[18] Alves D S M 2021 *Phys. Rev. D* **103** 055018 (*Preprint* 2009.05578)
[19] Bordes J, Chan H M and Tsou S T 2019 *Int. J. Mod. Phys. A* **34** 1950140 (*Preprint* 1906.09229)
[20] Nam C H 2020 *Eur. Phys. J. C* **80** 231 (*Preprint* 1907.09819)
[21] Kirpichnikov D, Lyubovitskij V E and Zhevlakov A S 2020 *Phys. Rev. D* **102** 095024 (*Preprint* 2002.07496)
[22] Fayet P 2021 *Physical Review D* **103** 035034 (*Preprint* 2010.04673)

[23] Hayes A C, Friar J L, Hale G and Garvey G 2021 (*Preprint* 2106.06834)
[24] Wong C Y 2022 QED Meson Description of the X17 and Other Anomalous Particles *Shedding light on X17* (*Preprint* 2201.09764)
[25] Baldini A M *et al.* (MEG II) 2018 *Eur. Phys. J. C* **78** 380 (*Preprint* 1801.04688)
[26] Balewski J *et al.* 2014 The DarkLight Experiment: A Precision Search for New Physics at Low Energies (*Preprint* 1412.4717)
[27] Ahdida C *et al.* (SHiP) 2020 (*Preprint* 2010.11057)
[28] G Azuelos t P The Montreal X-17 Project *New Scientific Opportunities at the TRIUMF ARIEL e-linac*
[29] BH Dongwi t P Search for Light Neutral Bosons in The TREK/E36 Experiment at J-PARC *New Scientific Opportunities at the TRIUMF ARIEL e-linac*
[30] Kou E, Urquijo P, Altmannshofer W *et al.* (Belle-II) 2019 *Prog. Theor. Exp. Phys.* **2019** 123C01 [Erratum: PTEP 2020, 029201 (2020)] (*Preprint* 1808.10567)
[31] Banerjee D, Bernhard J, Burtsev V E, Chumakov A G, Cooke D, Crivelli P, Depero E, Dermenev A V, Donskov S V, Dusaev R R, Enik T, Charitonidis N, Feshchenko A, Frolov V N, Gardikiotis A, Gerassimov S G, Gninenko S N, Hösgen M, Jeckel M, Kachanov V A, Karneyeu A E, Kekelidze G, Ketzer B, Kirpichnikov D V, Kirsanov M M, Kolosov V N, Konorov I V, Kovalenko S G, Kramarenko V A, Kravchuk L V, Krasnikov N V, Kuleshov S V, Lyubovitskij V E, Lysan V, Matveev V A, Mikhailov Y V, Molina Bueno L, Peshekhonov D V, Polyakov V A, Radics B, Rojas R, Rubbia A, Samoylenko V D, Shchukin D, Tikhomirov V O, Tlisova I, Tlisov D A, Toropin A N, Trifonov A Y, Vasilishin B I, Vasquez Arenas G, Volkov P V, Volkov V Y and Ulloa P (The NA64 Collaboration) 2020 *Phys. Rev. D* **101**(7) 071101 URL https://link.aps.org/doi/10.1103/PhysRevD.101.071101
[32] Cisbani E, Colonna N, Finocchiaro P, Fiore S, Gervino G, Gustavino C, Massimi C, Mastinu P, Mazzone A, Renga F and Viviani M 2021 https://cds.cern.ch/record/2766541
[33] Kievsky A, Rosati S, Viviani M, Marcucci L and Girlanda L 2008 *J. Phys. G* **35** 063101 (*Preprint* 0805.4688)
[34] Viviani M, Girlanda L, Kievsky A and Marcucci L E 2020 *Phys. Rev. C* **102**(3) 034007 (*Preprint* 2003.14059) URL https://link.aps.org/doi/10.1103/PhysRevC.102.034007
[35] Marcucci L E, Dohet-Eraly J, Girlanda L, Gnech A, Kievsky A and Viviani M 2020 *Front. in Phys.* **8** 69 ISSN 2296-424X (*Preprint* 1912.09751) URL https://www.frontiersin.org/article/10.3389/fphy.2020.00069
[36] Entem D and Machleidt R 2003 *Phys. Rev. C* **68** 041001 (*Preprint* nucl-th/0304018)
[37] Machleidt R and Entem D R 2011 *Phys. Rept.* **503** 1–75 (*Preprint* 1105.2919)
[38] Epelbaum E, Nogga A, Gloeckle W, Kamada H, Meissner U G and Witala H 2002 *Phys. Rev. C* **66** 064001 (*Preprint* 0208023)
[39] Piarulli M, Girlanda L, Schiavilla R, Kievsky A, Lovato A, Marcucci L E, Pieper S C, Viviani M and Wiringa R B 2016 *Phys. Rev. C* **94** 054007 (*Preprint* 1606.06335)
[40] Piarulli M *et al.* 2018 *Phys. Rev. Lett.* **120** 052503 (*Preprint* 1707.02883)
[41] Pastore S, Schiavilla R and Goity J L 2008 *Phys. Rev. C* **78** 064002 (*Preprint* 0810.1941)
[42] Pastore S, Girlanda L, Schiavilla R, Viviani M and Wiringa R 2009 *Phys. Rev. C* **80** 034004 (*Preprint* 0906.1800)
[43] Pastore S, Girlanda L, Schiavilla R and Viviani M 2011 *Phys. Rev. C* **84** 024001 (*Preprint* 1106.4539)
[44] Piarulli M, Girlanda L, Marcucci L E, Pastore S, Schiavilla R and Viviani M 2013 *Phys. Rev. C* **87** 014006 (*Preprint* 1212.1105)
[45] Schiavilla R *et al.* 2019 *Phys. Rev. C* **99** 034005 (*Preprint* 1809.10180)
[46] Andreas S, Niebuhr C and Ringwald A 2012 *Phys. Rev. D* **86**(9) 095019 URL https://link.aps.org/doi/10.1103/PhysRevD.86.095019
[47] Riordan E M, Krasny M W, Lang K, de Barbaro P, Bodek A, Dasu S, Varelas N, Wang X, Arnold R, Benton D, Bosted P, Clogher L, Lung A, Rock S, Szalata Z, Filippone B W, Walker R C, Bjorken J D, Crisler M, Para A, Lambert J, Button-Shafer J, Debebe B, Frodyma M, Hicks R S, Peterson G A and Gearhart R 1987 *Phys. Rev. Lett.* **59**(7) 755–758 URL https://link.aps.org/doi/10.1103/PhysRevLett.59.755
[48] Anastasi A *et al.* 2015 *Phys. Lett. B* **750** 633–637 (*Preprint* 1509.00740)
[49] Viviani M, Filandri E, Girlanda L, Gustavino C, Kievsky A, Marcucci L E and Schiavilla R 2022 *Phys. Rev. C* **105** 014001 (*Preprint* 2104.07808)
[50] Tilley D R, Weller H R and Hale G M 1992 *Nucl. Phys. A* **541** 1–104
[51] Zhang X and Miller G A 2021 *Physics Letters B* **813** 136061 (*Preprint* 2008.11288)
[52] Aleksejevs A, Barkanova S, Kolomensky Y G and Sheff B 2021 (*Preprint* 2102.01127)
[53] Koch B 2021 *Nucl. Phys. A* **1008** 122143 (*Preprint* 2003.05722)

Searching for New Physics with DarkLight at the ARIEL Electron-Linac

The DarkLight Collaboration, E Cline[1], R Corliss[1], J C Bernauer[1,2], R Alarcon[3], R Baartman[4], S Benson[5], J Bessuille[6], D Ciarniello[4], A Christopher[7], A Colon[1], W Deconinck[8], K Dehmelt[1], A Deshpande[1], J Dilling[4], D H Dongwi[9], P Fisher[6], T Gautam[7], M Gericke[8], D Hasell[6], M Hasinoff[10], E Ihloff[6], R Johnston[6], R Kanungo[11], J Kelsey[6], O Kester[4], M Kohl[7], I Korover[6], R Laxdal[4], S Lee[6], X Li[6], C Ma[1], A Mahon[4], J W Martin[12], R Milner[6], M Moore[6], P Moran[6], J Nazeer[7], K Pachal[4], T Patel[7], T Planche[4], M Rathnayake[7], M Suresh[7], C Vidal[6], Y Wang[6], S Yen[4]

[1] Department of Physics and Astronomy, Stony Brook University, Nicolls Road, Stony Brook, NY, 11794, USA
[2] RIKEN BNL Research Center, Brookhaven National Laboratory, 20 Brookhaven Ave, Upton, NY, 11973, USA
[3] Department of Physics, Arizona State University, 550 E Tyler Drive, Tempe, AZ, 85287, USA
[4] TRIUMF, 4004 Wesbrook Mall, Vancouver, BC, V6T 2A3, CA
[5] Thomas Jefferson National Accelerator Facility, 12000 Jefferson Avenue, Newport News, Virginia, 23606, USA
[6] Laboratory for Nuclear Science, Massachusetts Institute of Technology, 77 Massachusetts Avenue, Cambridge, MA 02139, USA
[7] Department of Physics, Hampton University, 268 Emancipation Drive, Hampton, VA, 23668, USA
[8] Department of Physics and Astronomy, University of Manitoba, 30a Sifton Road, Winnipeg, MB, R3T 2N2, CA
[9] Lawrence Livermore National Laboratory, 7000 East Ave, Livermore, CA 94550, USA
[10] Department of Physics and Astronomy, University of British Columbia, 6224 Agricultural Road, Vancouver, BC V6T 1Z1, CA
[11] Department of Physics and Astronomy, Saint Mary's University, 923 Robie Street, Halifax, NS, B3H 3C3, CA
[12] Department of Physics, University of Winnipeg, 515 Portage Avenue, Winnipeg, MB, R3B 2E9, CA

E-mail: ethan.cline@stonybrook.edu

Abstract. The search for a dark photon holds considerable interest in the physics community. Such a force carrier would begin to illuminate the dark sector. Many experiments have searched for such a particle, but so far it has proven elusive. In recent years the concept of a low mass dark photon has gained popularity in the physics community. Of particular recent interest is the ^{8}Be and ^{4}He anomaly, which could be explained by a new fifth force carrier with a mass of 17 MeV/c^2. The proposed Darklight experiment would search for this potential low mass force carrier at ARIEL in the 10-20 MeV/c^2 e^+e^- invariant mass range. This proceeding will focus on the experimental design and physics case of the Darklight experiment.

1. Introduction

Recently, there has been a focus in both experimental and theoretical physics communities on a mediator of a new fifth force with mass lower than 1 GeV/c^2. In addition to cosmological motivations, observed anomalies in measurements involving the muon and nuclear transitions hint at this possibility. For example, the observed 4.2σ deviation between the measured and expected anomalous magnetic moment of the muon [1] can be explained by a fifth force mediator with mass in the range 10 to 100 MeV [2]. The search for a dark photon has been extensively covered through the study of π^0-decay and much of the parameter space of coupling and mass that corresponds to these anomalies is excluded at 2σ [3]. However, it is not required that a potential dark sector coupling must be directly proportional to the electric charges, so a more general fifth force can not yet be ruled out.

It is possible to adjust the quark couplings of a fifth force to satisfy existing constraints and still allow such a force acting via lepton coupling to be experimentally detected. In addition to the muon g-2, other recently-reported experimental signatures motivate focused searches for a fifth force carrier at low energies. A group studying the decays of excited states of ^{8}Be and ^{4}He to their ground state have found a 6.8σ anomaly in the opening angle and invariant mass distribution of e^+e^- pairs produced in these transitions [4, 5, 6] (The ATOMKI anomalies). While these discrepancies may be the result of experimental effects or unidentified nuclear transitions they are also consistent with the production of a new boson with a mass around 17 MeV/c^2 (the X17 particle).

New bosons that couple atomic electrons with neutrons in the nucleus are also implicated in atomic physics experiments. The effect of this new interaction on energy levels and transition frequencies could be detected through precision isotope shift measurements. In particular, the scaled isotope shifts of two different transitions should exhibit a linear relationship (the so-called *King plot*). A deviation from linearity can be evidence of a new force mediator. Such deviations at the 3σ level have been reported [7] in the isotope shifts for five Yb$^+$ isotopes on two narrow optical quadrupole transitions $^2S_{1/2} \rightarrow {}^2D_{3/2} \rightarrow {}^2D_{5/2}$.

Motivated by these developments, we have designed the DarkLight experiment to use the future 50 MeV electron beam from the e-linac driver at the Advanced Rare IsotopE Laboratory (ARIEL) to search for evidence of the reported ATOMKI anomaly in e^+e^- final-states.

1.1. Beyond the Standard Model Interpretations

It has been shown that different couplings to quark and lepton flavors could reconcile the ATOMKI anomalies; different u and d couplings that produce protophobic or nearly protophobic interactions would satisfy current limits [8]. In light of that observation, the X17 anomaly has been interpreted in various more specific theoretical models as a new particle, a Z', axion, or other light pseudoscalar [9, 10, 11, 12]. There are also several proposed explanations for the X17 particle within the standard model framework arising from higher order effects [13, 14, 15].

2. Experiment Design

The beam energy at the ARIEL e-linac is relatively low, currently 31 MeV, but will be upgraded to 50 MeV. The advantage of the low beam energy is the smaller boost given to the produced e^+e^- pairs from the decay of the X17 boson. The smaller boost corresponds to a large opening angle. The proposed experiment takes advantage of these larger angles. DarkLight will run with a two spectrometer setup, optimized for the 17 MeV/c^2 invariant mass region. The spectrometers will be placed asymmetrically around the fixed foil target located in the beam line. The proposed experiment will measure the process $e^- Ta \rightarrow e^- TaX \rightarrow e^- Ta\ (e^+e^-)$ as a resonant excess of e^+e^- pairs on top of the QED background at the invariant mass of the X17.

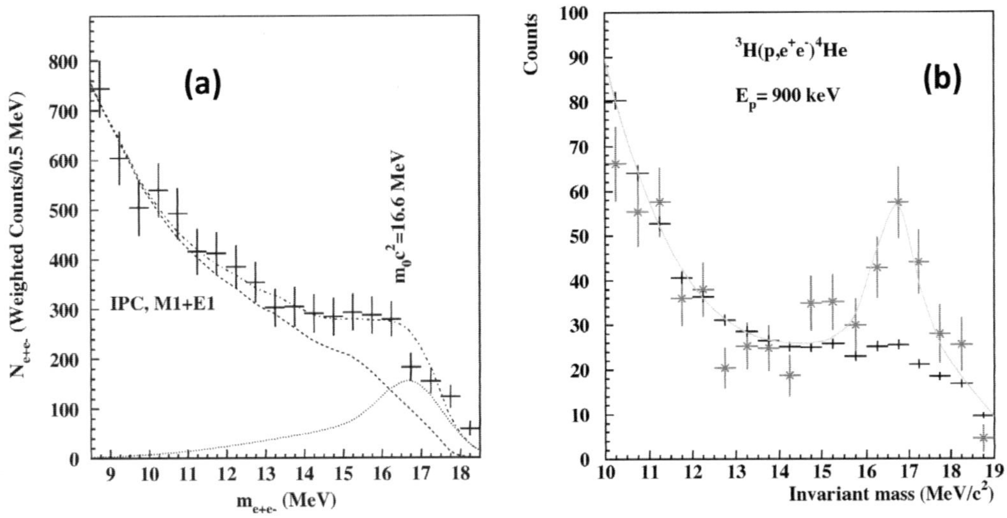

Figure 1. (a): Anomaly in ^{8}Be [4]. (b): Anomaly in ^{4}He [6].

2.1. The Electron Accelerator

TRIUMF's existing superconducting electron linac can currently produce an electron beam of up to 31 MeV in energy and peak intensities up to 3 mA. As a driver of ARIEL the e-linac is designed to deliver electrons to a photo-converter target station for the production of neutron-rich rare isotope beams via photo-fission. For testing and production running at 31 MeV, the experiment will be placed in front of the existing 10 kW beam dump (position A in Fig. 2). The linac will be operated in a continuous wave mode, with a bunch frequency of 650 MHz, and an average beam current of up to 300 μA. As this configuration involves minimal modifications to the beam line, data taking could commence in 2023.

The TRIUMF e-Linac is planning an upgrade to the overall beam energy deliverable to ARIEL. The approach to reach 50 MeV involves the installation of a second superconducting cavity and cryomodule. The cavity could be installed early/mid 2024, and production data taking for DarkLight at 50 MeV can commence late 2024.

A new 50 kW beam dump will be installed, and the experiment will move to Position B in Fig. 2. In order to facilitate simultaneous data taking with DarkLight and the rare isotope laboratory, a septum magnet and RF deflector will be added. The increased beam energy available will allow DarkLight to perform a comprehensive search of the parameter space of the X17.

2.2. Target

The experiment design assumes several beam energies, ranging from 30 MeV to 50 MeV with a current of 150 μA. It will impinge on a 1 μm tantalum foil. This produces an instantaneous luminosity of $\mathcal{L} = 5.2\,\mathrm{nb}^{-1}\,\mathrm{s}^{-1}$, and will cause a beam spread of approximately 0.5° downstream of the target.

The beam will heat up the foil with about 0.4 W, which can be dissipated via radiation for typical beam spot sizes. Also under consideration is a spinning foil disc target, linked to the accelerator Fast Shutdown, to protect against accidental melting.

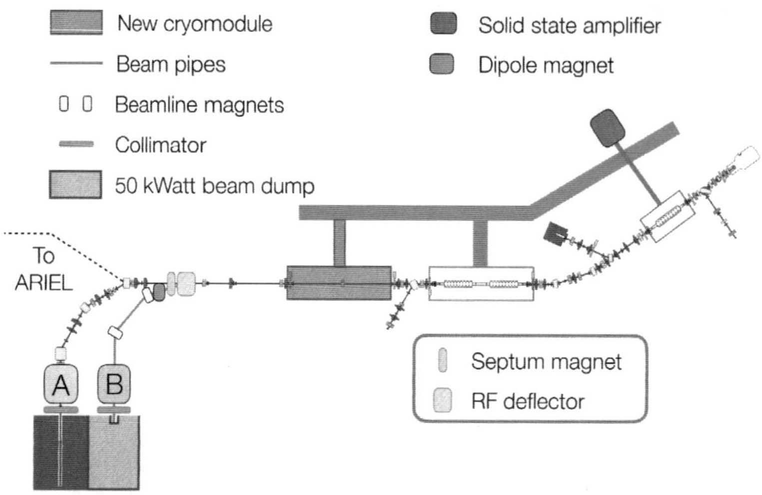

Figure 2. Floorplan of the ARIEL linac. For the first measurements at 31 MeV, the experiment will be placed in the area marked A. For 50 MeV beam, it will be moved to position B. The septum magnet and RF deflector enables concurrent beam delivery to ARIEL and DarkLight.

2.3. Spectrometers

The experiment will measure final state e^+e^- pairs using two dipole spectrometers, with very similar magnetic characteristics, to be built by MIT/Bates. The magnetic design and pole shapes of the spectrometer have been completed. Currently the mechanical design of the supports and coils are being finalized. Both spectrometers will be designed to nearly the same specifications, presented in Table 1.

Table 1. Design parameters for the spectrometers.

In-plane acceptance	$\pm 2°$
Out-of-plane acceptance	$\pm 5°$
Momentum acceptance	$\pm 20\,\%$
Minimum central angle	16°
Maximum central momentum	28 MeV
Dipole field	0.32 T
Nominal bend radius	30 cm
Pole gap	4 cm

The two spectrometers share a common design but will be operated at different currents to produce the desired magnetic fields. They are conventional iron-core magnets with simple, planar coils. The magnet design and pole face rotations were optimized for a 0.5 m distance from target to spectrometer entrance and for post-magnet trajectories suitable for tracking with 40 cm long GEMs. The final engineering of the magnet will include detailed design optimization to increase magnetic performance, minimize size, and maximize clearance to the exit beamline. The magnet in its present configuration weighs about 950 kg. The magnets will have full fiducialization to allow for laser tracking alignment and a six-strut mechanical support system to allow for 200 µm alignment (similar to other MIT-Bates designs). A 3D CAD rendering of the experiment and

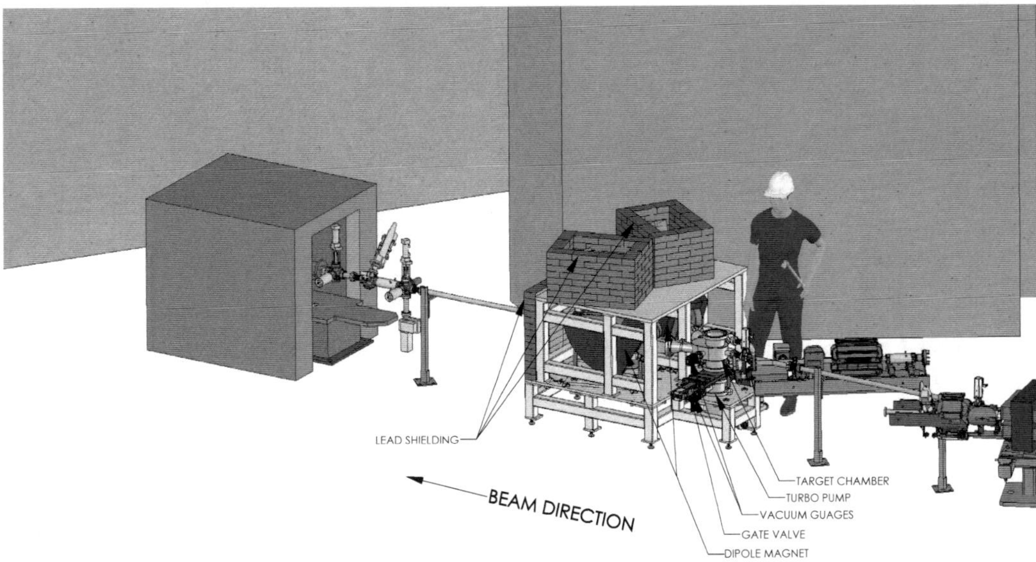

Figure 3. 3D CAD rendering of the conceptual design, with part of the shielding. Additional shielding around the target is anticipated. The exit beam line will be conical (6 cm radius at 3 m distance) to allow for the increased beam width from the target interaction.

spectrometers is shown in Fig. 3. The focal plane of each spectrometer will be instrumented with triple-GEM detector planes followed by trigger scintillators.

GEM detectors: Each spectrometer will be instrumented with an identical tracking detector system consisting of triple-GEM elements with an active area of 25x40 cm^2, which have been built by Hampton University. These modules have two-dimensional APV front-end readout cards with 400 μm pitch between strips. The APVs are read out into Multi-Purpose Digitzer front-end cards. They were constructed using the so-called NS2 scheme [16]. A similar system of these GEMs+APVs+MPDs has recently been mass-produced for the Super-Bigbite Spectrometer (SBS) construction at Jefferson Lab. The existing GEMs can be tested and commissioned with cosmics within 9-12 months.

Trigger Hodoscopes: The standard GEM readout requires a trigger signal. This will be generated from the coincidence of two fast trigger detectors in the spectrometers. To reduce accidental coincidences in the trigger logic, it is important to resolve the beam bunch clock of 650 MHz in the analysis. This high-resolution timing information must be provided by the trigger detector. When performing offline analysis, the timing can be corrected by the particle path length reconstructed from the tracking detector information. However, to reduce readout dead-time and data volume, it is important to be close to the ideal timing during data-taking. The main time dispersion is generated by the momentum-dependent dispersion inside the spectrometers. We therefore propose a trigger detector made from scintillator paddles, divided along the dispersive direction into 10 segments, each end read out by SiPMs.

The scintillator paddles will be made from a standard plastic scintillator material and have a size of about 150x30x2 mm^3.

3. Projected Reach

The shape of the background is dominated by the overall acceptance, so the irreducible background and random coincidences similar. To estimate the total background, we can scale the irreducible background up according to the predicted rates. The reach is calculated by integrating the background over the expected signal width ($\pm 1.7\sigma$), and calculating the fifth force coupling (ϵ^2) such that the signal would be bigger than a 2σ fluctuation of the background.

If ϵ^2 is small enough the signal may not be visually detectable on top of the background. Given the excellent statistical precision of the experimental background it will still be possible to detect fluctuations corresponding to a signal in the analysis.

Since random coincidences dominate the background, the pure random background will be very accurately measured by the experiment itself by mixing electron and positron spectrometer data from different events. This mixing destroys all correlations between the spectrometers, generating a pure sample of the random coincidences. Since, in principle, every combination of uncorrelated events can be used, the available statistics for the background measurement grows quadratically with the recorded number of events.

It is worth noting that at the kinematics and beam conditions of the experiment, the random coincidence background dominates, and scales with $\mathcal{L}^2$. The figure of merit (FOM) is given by the number of signal events divided by the square root of the background. This means in a regime where the random background dominates the FOM is independent of the instantaneous luminosity. The reach of the experiment depends only on the measurement time.

Figure 4 shows the achievable reach for the four settings assuming 1000 hours beam-time each. The precise optimization of the experimental reach is still in progress.

4. Summary

As described above, the DarkLight experiment is designed to seek the protophobic new force suggested by the ATOMKI anomalies. The experiment represents a natural evolution of a fixed-target experiment to reach the low mass expected if those anomalies truly represent a new particle. By using a high quality, low energy beam, the experiment simplifies the final states where only e^+e^- pairs, with no hadronic backgrounds detected. The experiment can operate at the instantaneous luminosity saturation point, where quadratic backgrounds dominate, and the FOM depends only on measurement time. The initial 1000 hour experimental run will seek a $13\,\mathrm{MeV}/c^2$ resonance and explore a small region of g-2 favored parameter space beginning in 2023, limited by the beam energy currently available at ARIEL. The apparatus will require only minor adjustment of the angles and magnet currents to accommodate the higher energies from the proposed ARIEL upgrades, which will enable searching in the $17\,\mathrm{MeV}/c^2$ range directly.

5. Acknowledgements

This work was supported by the National Science Foundation grant number PHY-2012114. We would also like to thank the Gordon and Betty Moore Foundation for their generous support of this workshop.

6. References

[1] B. Abi *et al.* (Muon g-2 Collaboration), Phys. Rev Lett **14**, 141801 (2021).
[2] M. Freytsis, G. Ovanesyan, and J. Thaler, Jour. High Energy Physics, **2010**, 111 (2010).
[3] D. Banerjee *et al.* [NA64], Phys. Rev. D **101**, 071101 (2020).
[4] A. Krasznahorkay *et al.*, Phys. Rev. Lett. **116**, 042501 (2016).
[5] A. Krasznahorkay *et al.*, Acta Physica Polonica B, **50**, 675.
[6] A. Krasznahorkay *et al.*, arXiv: 1910.10459, October 23, 2019.
[7] I. Counts *et al.*, arXiv:2004.11383, April 2020.
[8] J. Feng *et al.*, Phys. Rev. Lett. **117**, 071803 (2016).
[9] U. Ellwanger and S. Moretti, arXiv:1609.01669.

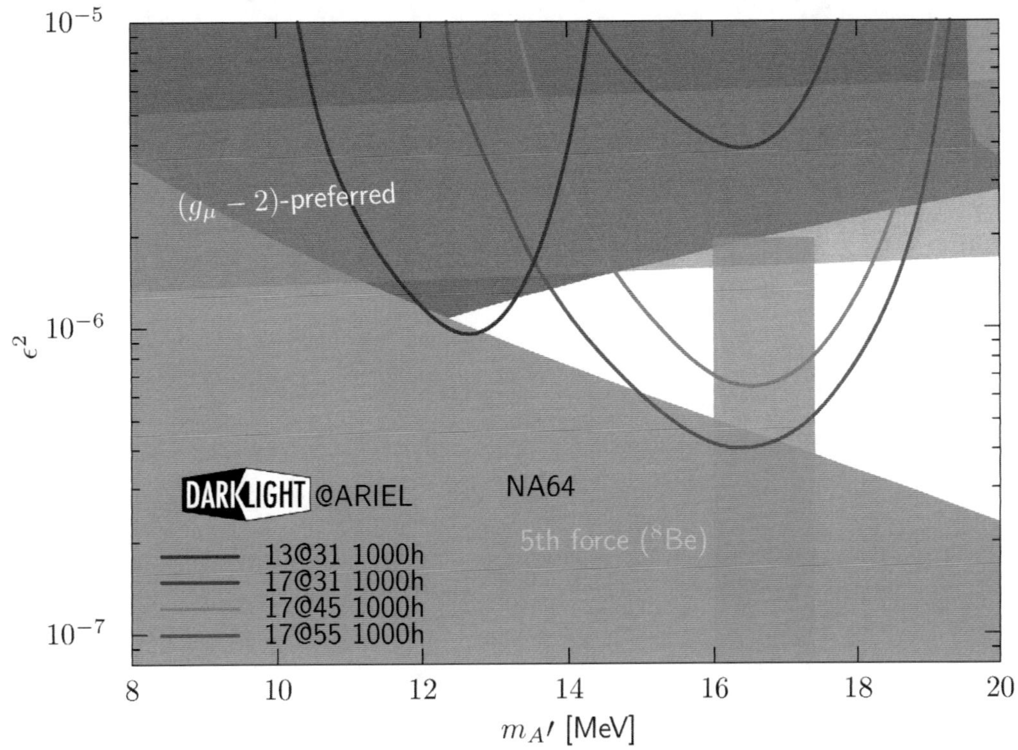

Figure 4. The projected reaches on a linear plot for three separate data taking runs: 13@31 (dark blue) – a 1000 h run at 31 MeV optimized for $m_{A'}$ = 13 MeV; 17@31 (light blue) – a 1000 h run at 31 MeV optimized for $m_{A'}$ = 17 MeV; 17@45 (light red) – a 1000 h run at 45 MeV optimized for $m_{A'}$ = 17 MeV; 17@55 (dark red) – a 1000 h run at 55 MeV optimized for $m_{A'}$ = 17 MeV. Light gray areas are excluded by other experiments sensitive to a lepton coupling. The dark gray area is excluded by electron g-2 only.

[10] D. S. M. Alves and N. Weiner, JHEP 1807, 092 (2017).
[11] L. Delle Rose *et al.* arXiv:1905.05031.
[12] L. Delle Rose *et al.* Front. in Phys. **7**, 73 (2019).
[13] X. Zhang and G.A. Miller, Phys. Lett. B**773**, 159 (2017).
[14] A. Aleksejevs *et al.*, arXiv:2102.01127.
[15] P. Kalman and T. Kaszthelyi, arXiv:2005.10643.
[16] D. Abbaneo *et al* 2014 *JINST* **9** C01053

Searches for dark photons at accelerators

Christopher Hearty
Department of Physics and Astronomy
University of British Columbia
Vancouver, BC, Canada
and
Institute of Particle Physics
Victoria, BC, Canada

E-mail: hearty@physics.ubc.ca

Abstract. Dark sector models attempt to describe observed dark matter properties, as well as other anomalies such as muon $(g-2)$, possible violations of lepton universality in B decay, and the possible observation of a new particle by the ATOMKI experiment. There is an extensive program of operating and proposed experiments to search for the new, low-mass, feebly-interacting particles predicted by such theories. I will focus on current searches at accelerators for dark photons, a component of many dark sector models. The results are interpreted in a basic model containing dark photons and dark matter.

1. Introduction: dark sector / dark matter / dark photons

A natural explanation of the observed astronomical dark matter properties is "freeze in". In the early universe, dark matter and standard model matter are in thermal equilibrium. Dark matter is created by the annihilation of standard model particles, and visa versa. As the universe expands and cools, dark matter production and annihilation essentially stop, resulting in the observed relic density. For low mass dark matter, this necessarily requires a new, feeble, interaction with standard model matter. Models containing such new, feeble interactions and low-mass particles are generically called the "dark sector".

A particularly simple model contains a dark matter particle χ and a dark photon A'. The dark photon has kinetic mixing of strength ε with the standard model photon. The dark photon decays and interactions are then related to the standard model.

The relative mass of the dark photon and the dark matter determine how the dark matter annihilates in the early universe. If the dark photon is heavier than the dark matter, the relic density depends on the parameter $y = \varepsilon^2\alpha_D(m_\chi/m_{A'})^4$, where α_D is the coupling between the dark photon and the dark matter; there are specific combinations of y and m_χ that match the observed value.

The relative masses also determine how the dark photon decays. If $m_{A'} > 2m_\chi$, the dark photon will decay to a pair of dark matter particles. The corresponding experimental signature is missing energy or mass, or scattering in a downstream neutrino-type detector. Otherwise, the dark photon decays like a virtual photon of mass $m_{A'}$. Experimentally, decays to a pair of leptons are typically (but not exclusively) used. For short dark photon lifetimes (large ε), the experimental signature is a narrow mass peak on a large standard model background from $\gamma^* \to \ell^+\ell^-$. For slightly longer lifetimes (smaller ε), the decay produces a displaced vertex, significantly reducing backgrounds. For even longer lifetimes, extensive shielding can strongly suppress standard model backgrounds.

This presentation is focused on operating experiments searching for dark photons decaying invisibly or to leptons, in the mass range $10\,\mathrm{MeV}/c^2 < m_{A'} < 10\,\mathrm{GeV}/c^2$. Even within this limited scope, it is not complete. For a recent comprehensive review of current and proposed experiments searching for a wide range of dark sector signatures, see Ref. [1] and the "Big Idea" white papers referenced therein. Much of the material in these proceedings is based on one of these white papers, Ref. [2], and on Ref. [3].

2. NA64

NA64 is a electron beam fixed target experiment located on the CERN SPS H4 beam line. It searches for a variety of new low-mass particles produced in 100 GeV or 150 GeV electron beams incident on a thick active target. The two dominant mechanisms for dark photon production are shown in Fig. 1. The primary mechanism is Bremsstrahlung, where the electron could be the incident beam particle, or an electron or positron produced in the electromagnetic shower in the target. There is also resonant production, which occurs when the center-of-mass energy of the collision between a positron and an atomic electron is equal to the dark photon mass. NA64 searches for invisible and visible decays of the dark photon with different detector configurations. They have also interpreted their data sets to set limits on other beyond-the-standard model particles, including axion-like particles [4], and $L_\mu - L_\tau$ Z' bosons [5].

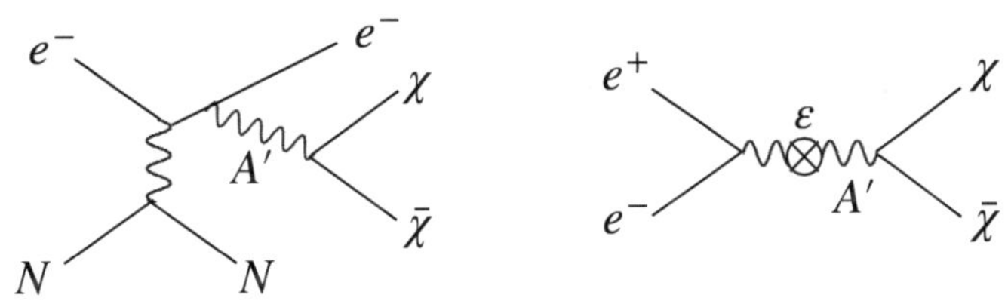

Figure 1. The primary production mechanisms for dark photon production in NA64 are (a) Bremsstrahlung from a beam or shower electron, and (b) resonant production from a positron produced in the electromagnetic shower in the active target. The same mechanisms produce the dark photons decaying to a visible (e^+e^-) final state.

2.1. NA64: invisible decays of the dark photon

NA64 collected 2.84×10^{11} 100 GeV electrons on target in 2016, 2017, and 2018, with the detector in the configuration for invisible decay search (Fig. 2) [6]. Additional data with this configuration was recorded in 2022. 100 GeV electrons are selected by two dipole magnets and beam-defining scintillators, are tracked by a set of micromegas chambers, (labeled MM in the diagram), and strike the 40 radiation-length thick active target (ECAL). The scintillator VETO ensures that no charged particles escape the ECAL. Neutral hadrons are detected in the three hadronic calorimeter modules HCAL.

In almost all standard model processes, the sum of the ECAL and HCAL energies will be consistent with the energy of the incident electron. An exception is dimuon production events, which deposit a range of energies in the ECAL and approximately 10 GeV in the HCAL, and are used to verify the accuracy of the detector simulation. After all selection criteria, no events are observed in the signal region, defined by low energy in the ECAL and no energy in the HCAL (Fig. 3a). A background of 0.5 events is expected from electron hadronic interactions in the beam line. The resulting limits on dark photon parameters are the world's best for dark photon masses below $0.3\,\mathrm{GeV}/c^2$ (Fig. 3b) [7].

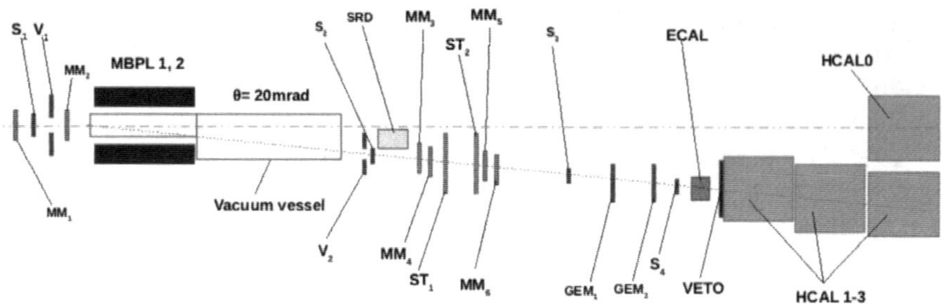

Figure 2. NA64 detector configuration for invisible dark photon searches [6].

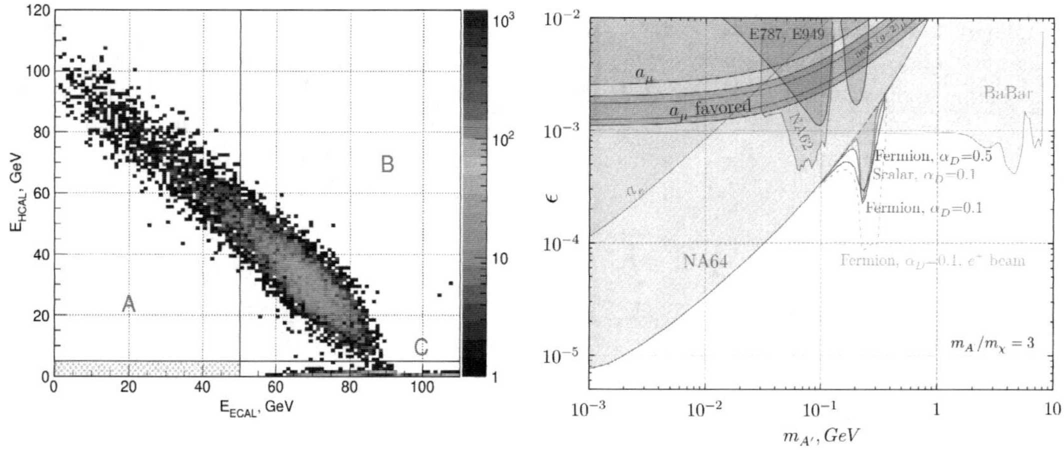

Figure 3. (a) Events observed by NA64 in the ECAL energy vs HCAL energy plane. No events are in the signal region (shaded) [6]. (b) The resulting limits are the world's best below 0.3 GeV/c^2 [7]. The limits from resonant production depend on α_D, the assumed coupling between the dark photon and dark matter.

2.2. *NA64: search for a dark photon (X17) decaying to e^+e^-*

The search for a dark photon (or X17 candidate) decaying to e^+e^- is performed with the NA64 detector in the visible mode configuration (Fig. 4). An addition tungsten/scintillator calorimeter WCAL is added upstream of the existing ECAL. The dark photon is produced in the WCAL, along with a secondary electromagnetic shower, then is required to have a sufficiently long lifetime to pass through the veto counter V2 before decaying the vacuum decay pipe. Signal events have no signal in V2, a signal consistent with two charged particles in scintillator S4 downstream of the decay pipe, and a total energy in WCAL plus ECAL consistent with the incident particle energy.

In this configuration, NA64 recorded 5×10^{10} 100 GeV electrons on target in 2017 and 3×10^{10} 150 GeV electrons on target in 2018. No signal candidates were observed, with an expected background of 0.08 events from $K_s^0 \to \pi^0\pi^0$ decays. The resulting 90% confidence level limits in the kinetic mixing parameter ε / dark photon mass plane are shown in Fig. 5 [8]. For a 16.7 MeV/c^2 mass, the excluded region is $1.2 \times 10^{-4} < \varepsilon < 6.8 \times 10^{-4}$. The region consistent with the ATOMKI anomaly, marked in red, is $2.0 \times 10^{-4} < \varepsilon < 1.4 \times 10^{-3}$.

To improve the sensitivity for larger values of ε, which correspond to shorter lifetimes, NA64

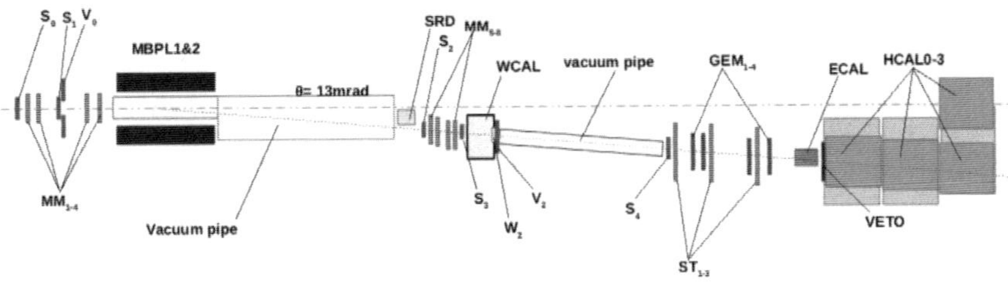

Figure 4. The visible configuration of NA64 adds an active target (WCAL), plus a decay volume, upstream of the existing ECAL [8].

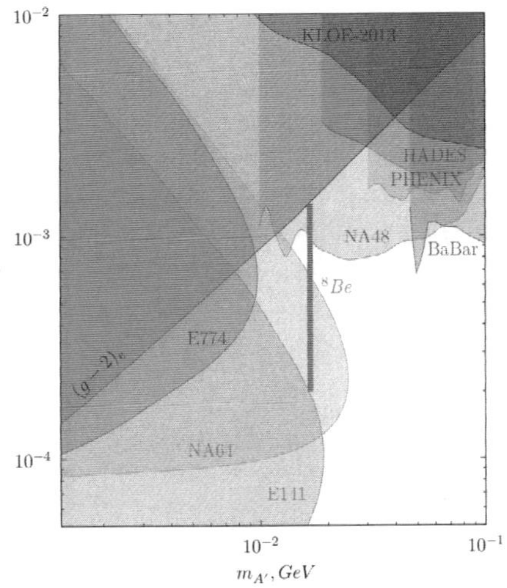

Figure 5. 90% confidence level limits on dark photon parameters from NA64 searches in the displaced e^+e^- final state. The region of parameter space consistent with the ATOMKI anomaly is marked in red [8].

is building a new, thinner WCAL, and will boost the beam energy to 200 GeV. They will also add a new magnetic spectrometer to measure the momentum of the decay electron and position. The next run in this configuration is planned for 2023. An estimated 7×10^{11} electrons on target will be needed to fully cover the ATOMKI parameter space [9].

3. Belle II

The Belle II experiment is located at the SuperKEKB e^+e^- collider in Tsukuba, Japan. It has an extensive program of B, charm, τ, and e^+e^- continuum physics aimed at detecting new physics through precision comparisons with standard model predictions [10]. These can include rare or forbidden decays, measurements to address the leptonic anomalies observed by the LHCb, *BABAR*, and Belle experiments, CP violation, and precision measurement of hadronic cross sections that are an essential input into predictions of the muon anomalous magnetic moment.

Belle II also has an wide program of searches for the direct production of new, low-mass particles—so called Dark Sector searches. These rely on the clean e^+e^- collider environment; the good solid angle coverage of the detector; and an extensive array of triggers for low multiplicity final states, include single photon and single track. The latter point is critical: the absence of such triggers has prevented Belle from undertaking a variety of important searches.

Dark photons are produced via initial state radiation, and then decay either invisibly or to a pair of leptons or other standard model particles (Fig. 6). Other cases are also studied, such as indirect dark matter [11], dark showers [12], or production in association with a dark Higgs [13], but are not discussed here.

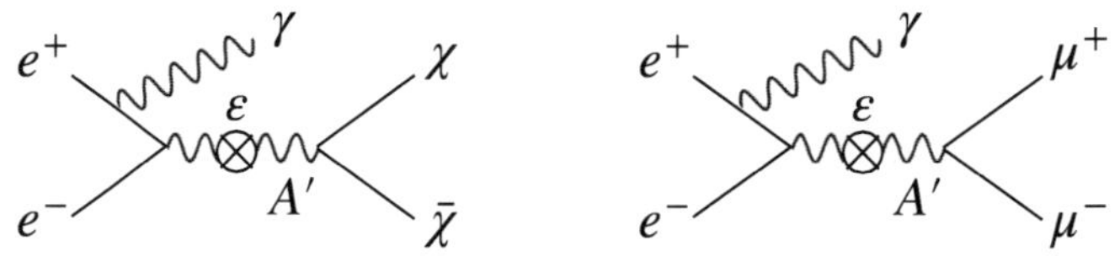

Figure 6. Feynman diagrams for the production of a dark photons at Belle II, followed by the decay to (a) dark matter or (b) a muon pair, as an example of a standard model decay.

Belle II is near-total upgrade of the original Belle experiment [14]. It has recorded $428\,\mathrm{fb}^{-1}$ since data taking started in March 2019, close to that recorded by *BABAR*, and is currently in long shutdown 1, during which they will install a new two-layer vertex detector. The design integrated luminosity of $50\,\mathrm{ab}^{-1}$ should be recorded by the early 2030's.

3.1. Belle II: invisible decays of the dark photon

The visible final state in this case is a single photon. Although its decay products are not reconstructed, the mass of the dark photon $m_{A'}$ is directly related to the center-of-mass energy of the photon E^*_γ under the assumption that there are no additional initial-state radiation photons in the event: $E^*_\gamma = E^*_B - m^2_{A'}/4E^*_B$, where E^*_B is the center-of-mass beam energy. A signal will therefore appear as a peak in the E^*_γ spectrum, with the expected distribution in the polar angle θ.

Backgrounds include:

- $e^+e^- \to \gamma\gamma$ in which one of the photons passes through the calorimeter without being detected, and then does not give a significant signal in the muon system. Particular care needs to be taken with this background, because the photon energy is peaked, which could give the appearance of a low-mass dark photon.
- $e^+e^- \to \gamma\gamma\gamma$ in which an initial-state radiation photon is lost down the beam beam, and a second photon hits the gap between the barrel and endcap calorimeters.
- $e^+e^- \to \gamma e^+e^-$, dominant at low photon energies, when both electrons are out of the acceptance.
- Cosmic rays.
- Non-luminosity single-beam backgrounds.

The analysis relies on quantifying each of these backgrounds in the E^*_γ vs. θ plane. This is in contrast to the *BABAR* analysis [15], which at high photon energy, did not attempt to predict the background level, and was therefore not able to detect a signal (but was able to set a limit).

Figure 7a shows the projected sensitivity in the kinetic mixing vs. dark photon mass plane. Belle II, in its initial analysis, will have sensitivity to regions of the parameter space that would correspond to that expected for the observed astronomical dark matter.

3.2. Belle II: visible decays of the dark photon

For visible decays of the dark photon, the final state consists of a photon and a pair of leptons. The initial search at Belle II will reconstruct all three particles. The signal consists of a narrow resonance in the $\ell^+\ell^-$ pair on top of a large but generally smooth standard model background. The process $e^+e^- \to \gamma J/\psi$ is an example of a standard model process that gives an identical signature. The J/ψ and other mass regions are excluded from the dark photon search.

Belle II has a considerably larger drift chamber than *BABAR*, which will give better mass resolution. The Belle II projection [10], shown in Fig. 7b, is derived from the *BABAR* limits [16], assuming a factor of two improvement in resolution.

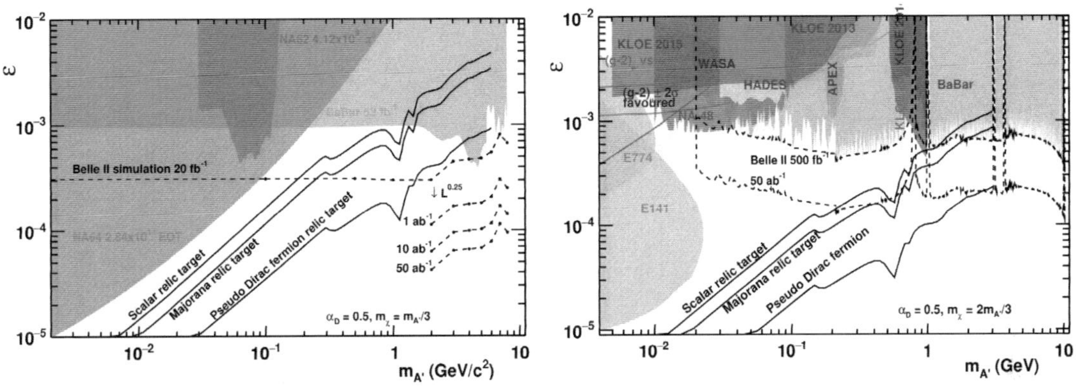

Figure 7. Projected Belle II sensitivity for (a) invisible dark photon decay and (b) visible dark photon searches in the kinetic mixing parameter ε versus dark photon mass plane. Adapted from [10]. Also shown are the dark matter relic density targets. In (b), this assumes that the dark matter particle χ is lighter than the dark photon, but sufficiently heavy that the dark photon cannot decay to two of them. There is no projection at higher luminosities for the invisible decay of low mass dark photons due to uncertainties in the cosmic ray background.

The *BABAR* analysis did not set limits below 20 MeV/c^2. The event generator used by *BABAR* to simulate the background from $e^+e^- \to \gamma e^+e^-$ did not correctly describe this kinematic region, and there is a large background from $e^+e^- \to \gamma\gamma$ in which one of the photons converts to e^+e^- in the beam pipe or vertex tracker material. Given the importance of investigating the ATOMKI anomaly, Belle II is undertaking an analysis focused on this mass region. It is also considering the statistically independent sample in which the initial-state-radiation photon is at low angles, and only the lepton pair is reconstructed. A larger data set will enable a displaced vertex search, with significant reach [17].

4. LHCb

LHCb is located at the pp collider LHC. Its unique forward spectrometer design gives it better physics reach than ATLAS or CMS for B and charm physics, despite operating at a lower luminosity. Dark photons are produced in pp collisions in Drell-Yan production, $q\bar{q} \to A'$, and in meson decay, primarily $\pi^0 \to \gamma A'$ and $\eta \to \gamma A'$.

LHCb has published a search for dark photons decaying to a muon pair, in both the prompt and displaced vertex configurations [18]. There is a large background in the prompt case, including a variety of standard model resonances that decay to $\mu^+\mu^-$, such as J/ψ and $\Upsilon(1S)$. These masses are excluded from the search region (Fig. 8). The displaced search also has non-negligible backgrounds from photon conversions to a muon pair, b hadron decays with two muons, and misidentified $K^0_S \to \pi^+\pi^-$ decays. The resulting displaced limits are unique; the prompt limits are the world's best for a range of masses.

LHCb has just undergone an extensive upgrade. As of fall 2022, it is commissioning the nearly-completed new detector towards data taking in 2023. Included in the upgrade will be a

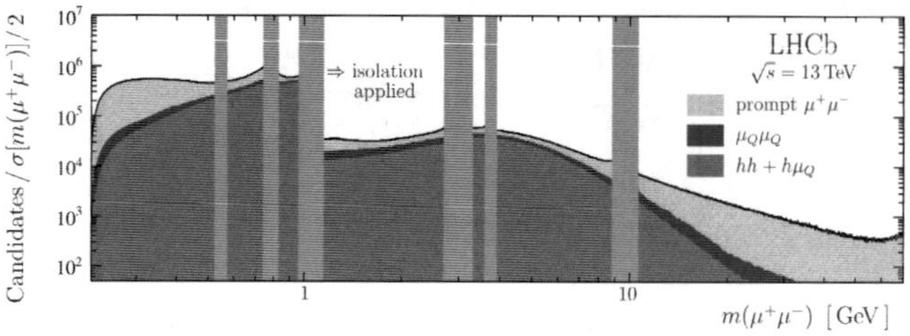

Figure 8. Mass spectrum of prompt $\mu^+\mu^-$ candidates. The coloured regions indicate the backgrounds from muons from heavy quark decays μ_Q and misidentified hadrons h. The grey bands reflect mass regions excluded due to mass peaks from standard model resonance decays [18].

trigger for inclusive electron pairs. The searches for dark photons decaying to electrons will be very powerful, covering a wide range of parameter space (Fig. 9) [19]. The limits from the muon pair analysis are the blue shaded regions.

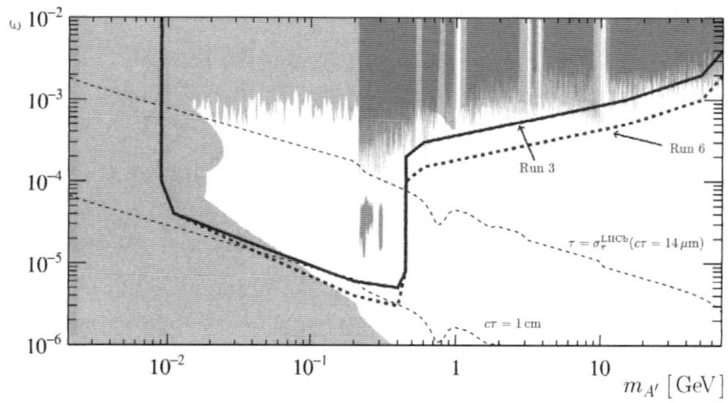

Figure 9. Projected reach for the upgraded LHCb detector, which will include triggers for inclusive e^+e^-. The sharp decrease in sensitivity above 400 MeV/c^2 is due to the loss of production from η decay and a K_s^0 veto [19].

5. FASER

FASER is a new LHC experiment, located 500 m downstream of ATLAS, on the axis of the *pp* collision. It is searching for dark photons and other dark sector particles, and includes an emulsion subdetector to reconstruct neutrino interactions. 100 m of rock and concrete shield the detector from all standard model collision products other than muons. FASER will search for dark photons produced by $\pi^0 \to \gamma A'$ and $\eta \to \gamma A'$. The dark photon will decay to e^+e^- in a decay volume after passing through a muon veto. The electrons will be tracked in three permanent magnet tracking stations, then deposit their energy in an electromagnetic calorimeter (Fig. 10).

FASER has started collecting data in fall 2022, and anticipates accumulating 150 fb^{-1} in LHC Run 3. This data set should enable data photon searches for masses up to 100 MeV/c^2 at small values of ε (Fig. 11) [20].

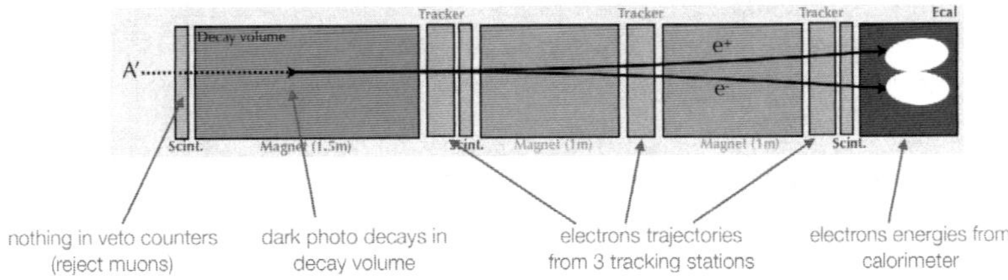

Figure 10. Diagram illustrating a dark photon search in FASER. Adapted from [20].

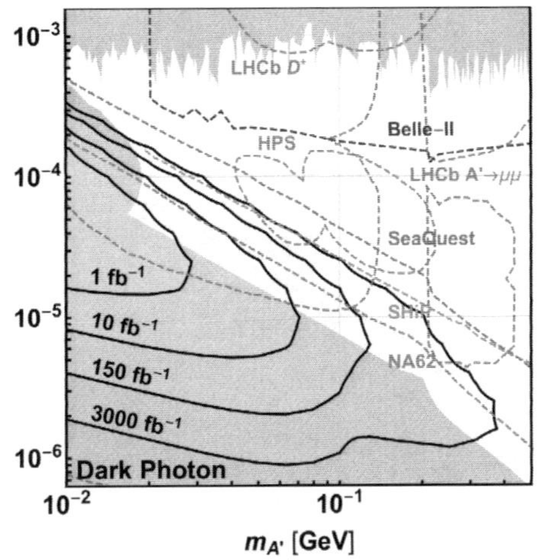

Figure 11. The projected FASER sensitivity to dark photon parameters for various data sets; approximately $150\,\mathrm{fb}^{-1}$ is expected in LHC Run 3. The sensitivity at large epsilon is set by the requirement that the dark photon have sufficient lifetime to each the detector. At small epsilon, sensitivity is limited by statistics [20].

6. COHERENT

COHERENT is located at the Oakridge National Research Laboratory spallation neutron source (SNS), which is also an excellent neutrino source. COHERENT is using a set of different detector technologies to study coherent elastic neutrino-nucleus scattering CENνS, in which an entire nucleus recoils elastically from a scattered neutrino [21].

Dark photons are by the SNS through the decay of π^0 and η mesons, and by the nuclear absorption of π^-. Dark matter particles produced in the decay of dark photons can similarly produce CENνS by the exchange of a virtual dark photon. The expected rate is small compared to that of neutrinos, but has a different time structure (Fig. 12) [22]. The neutrino signal has a prompt component from ν_μ produced in π^+ decay, and a component from ν_e and $\bar{\nu}_\mu$ that has the muon lifetime. The dark matter signature, on the other hand, is entirely prompt. Single-phase liquid argon, cryogenic CsI, and a high-purity Germanium array will collect data for the next several years, producing competitive results in the search for accelerator-produced dark matter [23]. An anticipated doubling of the SNS power by the end of the decade, and a possible second target station in the mid-2030's would significantly enhance the neutrino and dark sector program.

Journal of Physics: Conference Series **2391** (2022) 012011 doi:10.1088/1742-6596/2391/1/012011

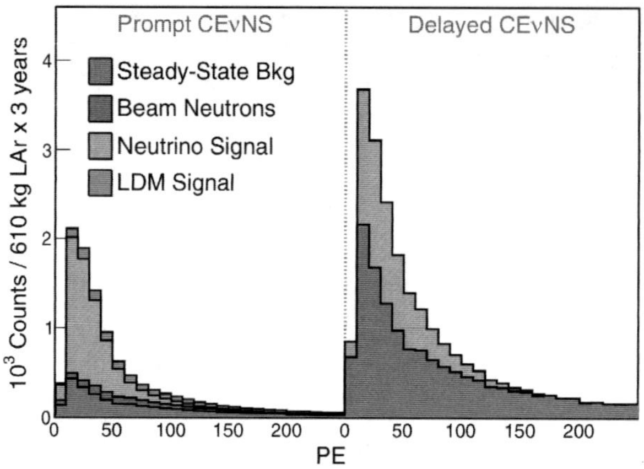

Figure 12. The scattered dark matter CENνS signal in COHERENT is entirely prompt, left, whereas the neutrino signal has both a prompt component, and component that follows the muon lifetime (right) [22].

7. Outlook

The next few years will see exciting new results from the experiments outlined above. Beyond that, there are numerous proposals for new projects. CERN is considering a new detector focused on dark sector physics for the SPS fixed target program, and is reviewing proposals for additional LHC experiments looking at long-lived particles. In the United States, the Snowmass planning process is drawing to a close. Two figures from the Snowmass report "Dark Sector Physics at High-Intensity Experiments" [1] illustrate the promising prospects (Fig. 13).

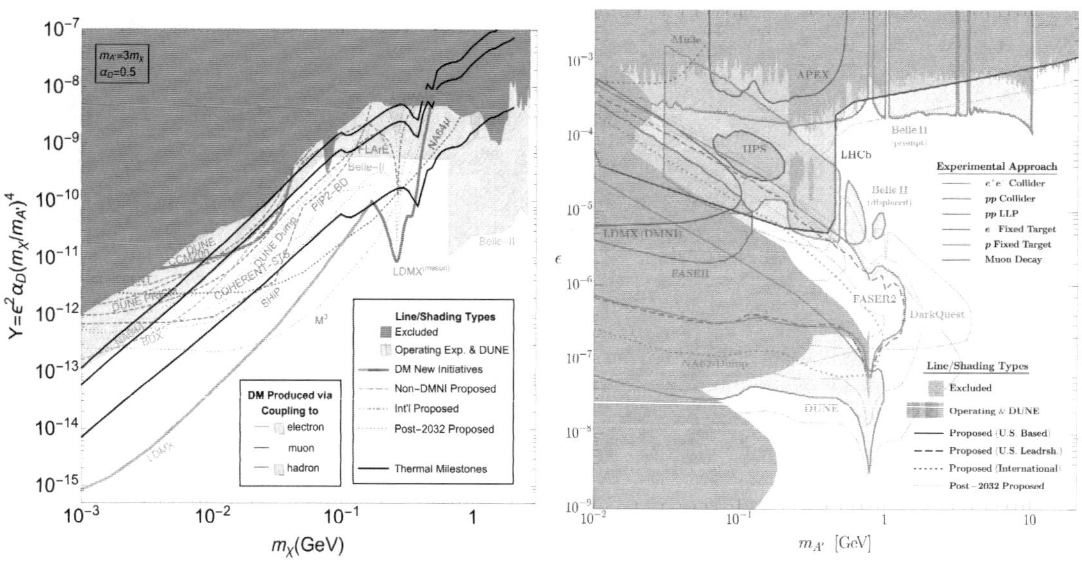

Figure 13. (a) Dark matter production at accelerators, and (b) visible dark-photon decays. Coloured shaded regions correspond to operating experiments and DUNE; proposed projects are indicated with unfilled lines [1].

Journal of Physics: Conference Series **2391** (2022) 012011 doi:10.1088/1742-6596/2391/1/012011

References

[1] Gori S *et al.* 2022 (*Preprint* 2209.04671)
[2] Batell B, Blinov N, Hearty C and McGehee R 2022 Exploring Dark Sector Portals with High Intensity Experiments *2022 Snowmass Summer Study* (*Preprint* 2207.06905)
[3] Graham M, Hearty C and Williams M 2021 *Ann. Rev. Nucl. Part. Sci.* **71** 37–58 (*Preprint* 2104.10280)
[4] Banerjee D *et al.* (NA64) 2020 *Phys. Rev. Lett.* **125** 081801 (*Preprint* 2005.02710)
[5] Andreev Y M *et al.* (NA64) 2022 *Phys. Rev. D* **106** 032015 (*Preprint* 2206.03101)
[6] Banerjee D *et al.* 2019 *Phys. Rev. Lett.* **123** 121801 (*Preprint* 1906.00176)
[7] Andreev Y M *et al.* 2021 *Phys. Rev. D* **104** L091701 (*Preprint* 2108.04195)
[8] Banerjee D *et al.* (NA64) 2020 *Phys. Rev. D* **101** 071101 (*Preprint* 1912.11389)
[9] Depero E 2021 X17 in the na64 experiment presented at Shedding light on X17, Rome, Italy https://agenda.infn.it/event/26303/
[10] Aggarwal L *et al.* (Belle II) 2022 (*Preprint* 2207.06307)
[11] Duerr M, Ferber T, Hearty C, Kahlhoefer F, Schmidt-Hoberg K and Tunney P 2020 *JHEP* **02** 039 (*Preprint* 1911.03176)
[12] Bernreuther E, Böse K, Ferber T, Hearty C, Kahlhoefer F, Morandini A and Schmidt-Hoberg K 2022 (*Preprint* 2203.08824)
[13] Abudinén F *et al.* (Belle II) 2022 (*Preprint* 2207.00509)
[14] Abe T *et al.* (Belle II) 2010 (*Preprint* 1011.0352)
[15] Lees J P *et al.* (BaBar) 2017 *Phys. Rev. Lett.* **119** 131804 (*Preprint* 1702.03327)
[16] Lees J P *et al.* (BaBar) 2014 *Phys. Rev. Lett.* **113** 201801 (*Preprint* 1406.2980)
[17] Ferber T, Garcia-Cely C and Schmidt-Hoberg K 2022 *Phys. Lett. B* **833** 137373 (*Preprint* 2202.03452)
[18] Aaij R *et al.* (LHCb) 2020 *Phys. Rev. Lett.* **124** 041801 (*Preprint* 1910.06926)
[19] Craik D, Ilten P, Johnson D and Williams M 2022 LHCb future dark-sector sensitivity projections for Snowmass 2021 *2022 Snowmass Summer Study* (*Preprint* 2203.07048)
[20] Petersen B 2019 Faser status and faserν proposal lHCC 140 open session https://indico.cern.ch/event/854465/
[21] Akimov D *et al.* (COHERENT) 2021 *Phys. Rev. Lett.* **126** 012002 (*Preprint* 2003.10630)
[22] Akimov D *et al.* (COHERENT) 2020 *Phys. Rev. D* **102** 052007 (*Preprint* 1911.06422)
[23] Akimov D *et al.* 2022 The COHERENT Experimental Program *2022 Snowmass Summer Study* (*Preprint* 2204.04575)

Search for Light Neutral Bosons in The TREK/E36 Experiment at J-PARC

Dongwi H. Dongwi (on behalf of the TREK/E36 Collaboration)
7000 East Ave., Livermore, CA 94550-9234
E-mail: dongwi1@llnl.gov

Abstract. The Standard Model (SM) represents our best description of the subatomic world and has been very successful in explaining how elementary particles interact under the influence of the fundamental forces. Despite its far reaching success in describing the building blocks of matter, the SM is still incomplete; falling short to explain dark matter, baryogenesis, neutrino masses and much more. The E36 experiment conducted at J-PARC in Japan, allows for sensitivity to search for light $U(1)$ gauge bosons, in the muonic K^+ decay channel. Such $U(1)$ bosons could be associated with dark matter or explain established muon-related anomalies such as the muon $g-2$ value, and perhaps the proton radius puzzle. A realistic simulation study was employed for these rare searches in a mass range of 20 MeV to 100 MeV. Preliminary results of the upper limits for the A' branching ratio $\mathcal{B}r(A')$ extracted at 95% CL, will be presented.

1. Introduction

The primary goal of the TREK/E36 experiment was to provide a high precision electroweak measurement in order to test lepton universality, which is expressed as an identical coupling constant of the charged lepton family (e, μ, and τ). Lepton universality is a staple of the Standard Model (SM) and any violation of this would be clear evidence of New Physics (NP) beyond the SM. The ratio of the two-body K^+ decay widths, of R_K is very precise because to a first approximation the strong interaction dynamics cancel, leaving behind equation 1 [1, 2]

$$\begin{aligned} R_K^{SM} &= \frac{\Gamma(K^+ \to e^+\nu)}{\Gamma(K^+ \to \mu^+\nu)} \\ &= \frac{m_e^2}{m_\mu^2}\left(\frac{m_K^2 - m_e^2}{m_K^2 - m_\mu^2}\right)^2 (1+\delta_r) \\ &= (2.477 \pm 0.001) \times 10^{-5} \end{aligned} \tag{1}$$

here δ_r represents radiative corrections, detailed calculations of which were carried out in [3]. Thus the SM value for R_K^{SM} has been calculated to high accuracy ($\Delta R_K/R_K \sim 0.4 \times 10^{-4}$) thereby making it possible to search for NP effects by conducting a precise measurement of R_K [1].

Additionally, the TREK/E36 experiment is sensitive to searches for light neutral particles in exotic kaon decay modes of $K^+ \to \mu^+\nu A'$ and/or $K^+ \to \pi^+ A'$, followed by a prompt decay of $A' \to e^+e^-$. Here, the A' channel of interest is in the muonic K_μ decay, $K^+ \to \mu^+\nu A'$. This

A', often dubbed a dark photon, could also be construed as a light neutral $U(1)$ boson, and might be a hidden force carrier of the dark sector associated with dark matter [4]. An A' that is weakly coupled to the SM and is sufficiently light could decay into observable di-lepton pairs, electron-positron pairs, which can be used to reconstruct its invariant mass. Furthermore, the A' could help resolve, simultaneously, the proton radius puzzle and anomalous magnetic moment of the muon $g_\mu - 2$ [5–8]. Such hypothetic particles can also be conceived without violating existing constraints if they are fine-tuned and non-universally coupled [6–11]. In this case there would be a prediction of a strong observable signal in kaon decays, in particular in the calculable leptonic radiative mode $K^+ \to \mu^+ \nu e^+ e^-$ [7,8].

1.1. *Carlson-Rislow Model*

One such model, that proposes observable signals in leptonic K^+ decays, is the Carlson-Rislow model [8], which introduces a pair of new particles with polar and axial vector couplings to protons and muons respectively, in order to explain the proton radius puzzle while also obeying the $(g-2)_\mu$ discrepancy. The coupling strength of polar vector was chosen to explain the proton radius puzzle. The axial vector coupling enters the expression for the anomalous magnetic moment with an opposite sign to the polar vector, and was fine tuned to explain the $(g-2)_\mu$ discrepancy [7]. Rare kaon decays provide a good laboratory to test NP explanations for the proton radius puzzle. Beranek and Vanderhaeghen [12] have calculated constraints that the dark photon via kinetic mixing as well as lepton universality violating couplings are expected to contribute to the $K^+ \to \mu^+ \nu e^+ e^-$ decay. Carlson and Rislow [7,8] calculated their light neutral boson's contribution to the $K^+ \to \mu^+ \nu e^+ e^-$ amplitude by modifying the photon's propagator and charged fermion couplings as follows [8]

$$\frac{-i}{q^2} \to \frac{-i}{q^2 - m_{A'}^2 + i m_{A'} \Gamma}, \tag{2}$$

$$-ie\gamma^\nu \to \gamma^\nu \varepsilon (C_V(m_{A'}) + C_A(m_{A'})\gamma^5), \tag{3}$$

where $C_V(m_{A'})$ and $C_A(m_{A'})$ are the polar and axial vector couplings to muons calculated in [7]. The couplings $C_{V,A}$ were chosen so as to obtain an extra 310 μeV muonic hydrogen Lamb shift from an electrophobic spin-0 or spin-1 particle as described in [7]. Figure 1 shows the muonic kaon decay involving an A' $K^+ \to \mu^+ \nu A'$ followed promptly by $A' \to e^+ e^-$.

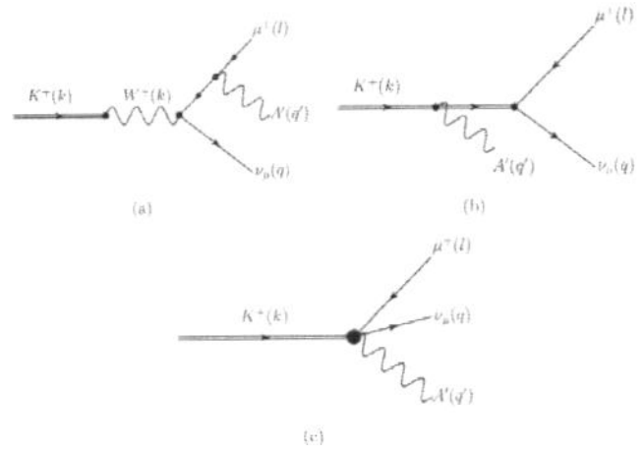

Figure 1: A' production amplitude. Feynman diagram showing the channel of interest $K^+ \to \mu^+ \nu_\mu A'$ [12].

Carlson and Rislow have allowed for the possibility that the new particle has scalar and pseudoscalar couplings to muons and protons. They have modeled its contribution to the $K^+ \to \mu^+ \nu e^+ e^-$ amplitude by modifying the photons propagator as before and modifying the charged fermion couplings as [7]

$$- ie\gamma^{\mu} \rightarrow -i\varepsilon(C_S(m_{A'})) + iC_P(m_{A'}\gamma^5), \quad (4)$$

where $C_S(m_{A'})$ and $C_P(m_{A'})$ are the scalar and pseudoscalar couplings to muons as calculated in [7]. They have also calculated mass limits on polar and axial vector (as well as scalar and pseudoscalar) particles due to constraints placed by searches of muonic kaon decays $K^+ \rightarrow \mu X$. Their results are presented in figure 2. The solid curve in both figure 2 (a) and (b) is the result for a single particle with both polar (scalar) and axial vector (pseudoscalar) couplings. The horizontal line indicates the experimental limit, and ϕ can be taken to represent the A'.

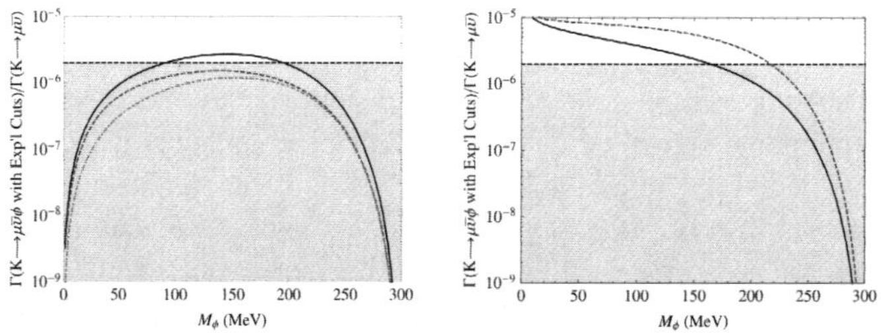

(a) (b)

Figure 2: Mass limits on particles with scalar and pseudoscalar (a), and polar and axial vector (b) couplings due to constraints placed by $K^+ \rightarrow \mu X$ searches. The solid curve is the result for a single particle with both polar (scalar) and axial vector (pseudoscalar) couplings. In (a) the contributions of scalar coupling is represented by the dashed curve and the dash-dotted curve represents the pseudoscalar coupling. In (b) the dashed curve is the result for separate polar and axial vector particles with equal masses, and in both (a) and (b) the green reqion represents the experimental limit (horizontal line) [7].

Carlson and Rislow have calculated the branching ratio $\mathcal{B}r(K^+ \rightarrow \mu^+\nu A')$ to be $\mathcal{O}(-5)$ for polar and axial vector couplings. Their prediction indicate that a strong signal peak over the SM background would be observed experimentally as shown in figure 3.

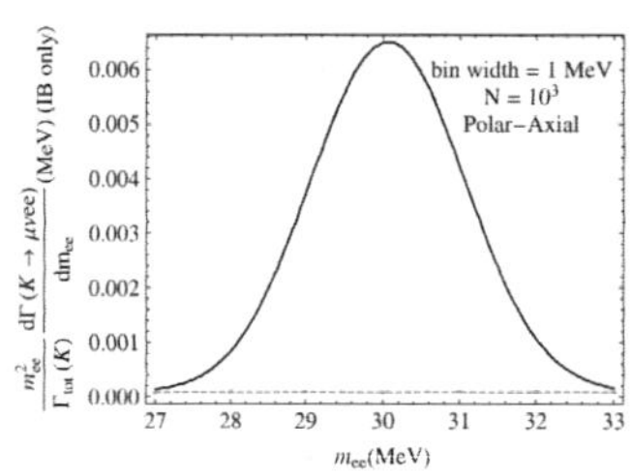

Figure 3: Strong signal to background prediction. The QED background prediction for $K^+ \rightarrow \mu^+\nu e^+e^-$ (red dashed curve) and the prediction with a 30 MeV lepton universality violating A' of the Carlson model [11].

Given the strong signal to background expectation of the Carlson model, the TREK/E36 experiment which is sensitive to light neutral bosons below 300 MeV would be able to test the efficacy of the Carlson model. The search would include a charged μ^+ that is detected in the spectrometer gap and two clusters in the CsI barrel from e^+-e^- pairs, from which a peak search will be conducted in the invariant mass spectrum M_{ee}.

2. Light Neutral Boson Search with the TREK/E36 Detecor Apparatus

TREK/E36 was set up and fully commissioned at the K1.1BR kaon beamline between the fall 2014 and spring 2015. Production data for the TREK/E36 experiment was completed by the end of 2015, which made use of an upgraded version of the KEK-PS E246 12-sector superconducting toroidal spectrometer used in a previous T-violation experiment via transverse muon polarization of muons in $K^+ \to \mu^+\nu\pi^0$ ($K_{\mu3}$) decays at KEK [2, 13]. The schematic longitudinal and end views of the detector system are shown in figure 4.

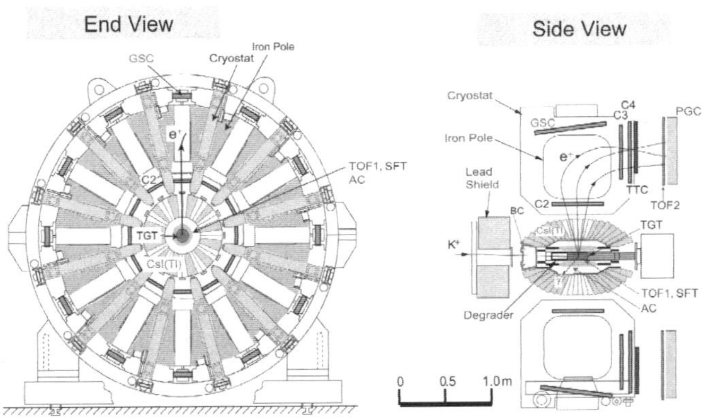

Figure 4: TREK/E36 apparatus. General cross-sectional and longitudinal views of the detector system.

The TREK/E36 apparatus upgrade from E-246 consisted of:

- scintillating fiber target used to stop the incoming kaon beam had a smaller diameter to minimize multiple scattering and energy loss of the outgoing decay particles [2]
- addition of Spiral Fiber Tracker [2, 14]
- redundancy in particle ID systems used to distinguish between e and μ with high efficiency and low misidentification probability [2]
- faster readout electronics of the CsI(Tl) calorimeter with a pile-up capable data acquisition system using FPGA based waveform digitization [2, 15].

The incoming K^+ was tagged with a Fitch Čerenkov detector [2], in order to distinguish it from a π^+, before stopping and decaying at rest in the active target, a matrix of 256 scintillating fibers oriented longitudinally along the beam, which determines the location of the kaon stop in the transverse plane. Surrounding the target was the a Spiral Fiber Tracker (SFT), which consisted of two pairs of fiber layers spiraling in *left/right* helicity and provided the z-coordinate of the of the outgoing decay particle(s) [14]. The target-SFT system was surrounded by 12 time-of-flight counter (TOF1) and 12 aerogel Čerenkov counters (AC) aligned with the 12 sectors of the toroidal spectrometer. This entire inner system is called the *Central Detector* system (CD).

A highly segmented and large acceptance CsI(Tl) calorimeter barrel which consisted of 768 crystal and covered about 75% of 4π [16] was used to identify and correct for structure dependent (SD) background events and also search for light neutral bosons A'. The calorimeter featured 12 holes (known as *muon holes*) aligned with the sectors of the spectrometer allowing for charged particles such as e^+, μ^+ and π^+ to be tracked through 1 of the 12 toroidal magnet gaps, and momentum analyzed with Multi-Wire Proportional Chambers (MWPC) C2, C3 and C4. The A' search necessitated a charged particle tracked in the gap that has a vertex originating within the fiducial volume of the target (*good gap event*) and 2-clusters in the CsI calorimeter along with at least 3 TOF1 counters that registered a charged particle (ideally: e^+, e^- and a good gap μ^+). At the exit of each magnet gap were sets of fast scintillator detectors (TTC and TOF2) and lead glass counters (PGCs) [17] that provided both trigger signals and e^+/μ^+ particle ID. The full suite of particle ID systems and tracking can also be deployed to suppress background events in the invariant mass spectrum when performing the A' search.

3. CsI Analysis and A' Search Strategy

Calibration of the CsI barrel was carried with monochromatic $K_{\mu 2}$ energy. A cross check of the calibration was performed with the $K_{\pi 2}$ momentum pre-selection, and the associated 2-γ clusters as shown in figure 5.

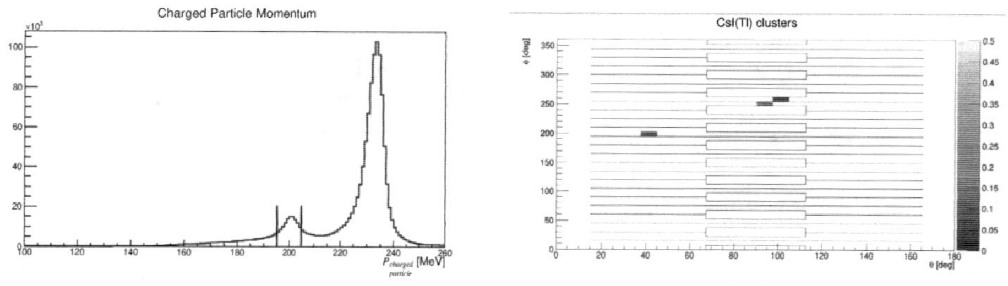

(a) (b)

Figure 5: Pre-selected $K_{\pi 2}$ cluster hits in the CsI calorimeter. $K_{\pi 2}$ momentum selection cut in (a), with black lines indicating the generous selection invterval. The CsI event viewer (b) showing a given cluster hit pattern.

Two photon cluster events in the CsI detector were used to reconstruct the invariant mass of the π^0, well as the total cluster energy. An energy-weighted centroid technique was employed for this reconstruction. Since the CsI(Tl) was operated with high hardware threshold ($\approx$ 20 MeV), only crystals with energies greater than 20 MeV participated for a given event. For this reason, single crystals were considered to be part of a cluster, namely a *single crystal cluster*.The reconstructed π^0 invariant mass from simulation (red) was observed to have a thinner peak compared to the data and a similar situation was also observed for the sum of the 2γ cluster energies, as seen in figure 6.

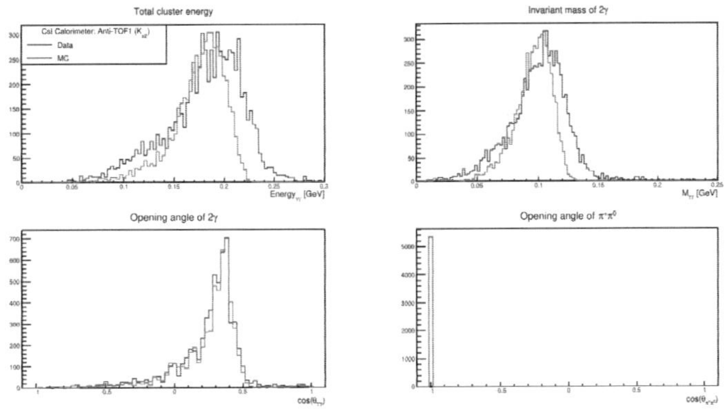

Figure 6: Plots of total energy, invariant mass of 2γ clusters, opening angle of 2γ's, $cos(\theta_{\gamma\gamma})$, and opening angle of $\pi^+\pi^0$ $cos(\theta_{\pi^+\pi^0}) \leq -.99$. Better agreement between data and simulation was achieved by requiring that only a single TOF1 has registered a hit per event.

3.1. A′ Search Strategy

A necessary condition for the A' search is to have a so-called good gap event. This means that a charged particle has escaped from the openings in the CsI barrel, and had its momentum evaluated by the spectrometer magnet, and has registered hits in the ToF and C-detector elements. In the TREK/E36 experiment, theA' search was performed in the muonic kaon decay channel $K^+ \rightarrow \mu^+\nu A'$, followed by the prompt $A' \rightarrow e^+e^-$ decay. The trigger condition for the A' required a charged particle in the gap and 3 TOF1 counters that registered a hit. Ideally, one of the three TOF1 counters corresponds a tracked μ^+ while the other two register e^+ and e^-. The criteria for clusters in the calorimeter can be summarized as follows:

(i) good gap event, i.e. a charged particle has a track that originated in the target and was tracked by the three MWPCs,

(ii) momentum cut $p < 230$ MeV$/c$,

(iii) only two clusters were observed in the calorimeter,

(iv) 3 TOF1 counters with registered hits

The e^+-e^- pair would then go on to form two clusters in CsI calorimeter, which can be used to produce an invariant mass spectrum M_{ee}. A peak search over a continuous background is needed to determine if an A' signal is present. With this criteria in place, the search for A' proceeds as laid out in figure 7.

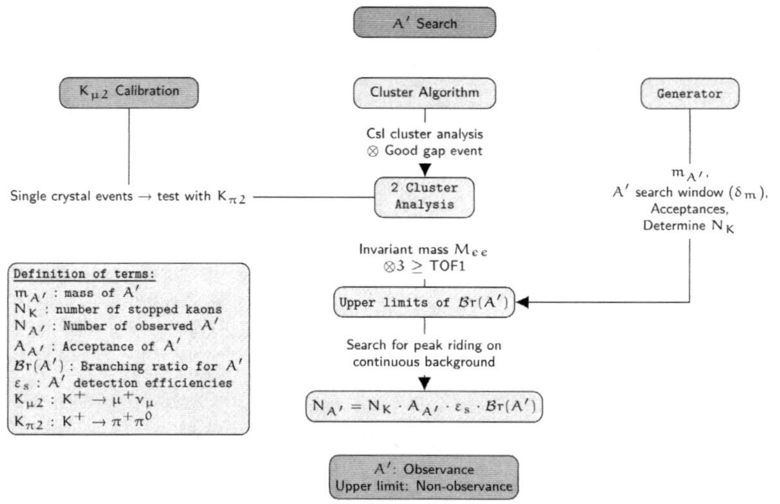

Figure 7: A' search strategy analysis flow schema, showing the combination of cluster analysis, *good gap* event, and upper limit at 95% CL.

4. Preliminary Upper Limit Extraction

Aforementioned, the A' signal search required a TOF1 multiplicity of 3 or greater, implying that three charged particles passed through the TOF1 counters. An additional cut on a good gap event where a charged particle registered hits in the tracking detectors C2, C3 and C4, was deployed provided that the track vertex lies within the fiducial volume of the target. A further endpoint momentum cut, shown in figure 8, was applied for the express purpose of suppressing the $K_{\mu 2}$ muons.

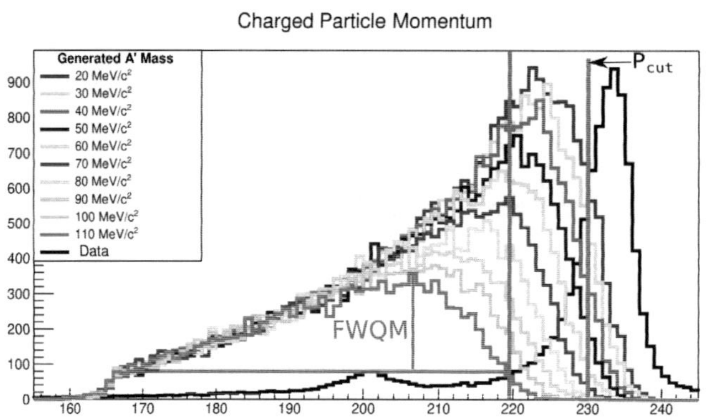

Figure 8: Endpoint momentum cut, with a hard cut placed at 230 MeV$/c$. A variable full-width at quarter max (FWQM) cut was employed iff less then the hard cut at 230 MeV$/c$.

Noticeable effects of these sequential background suppression cuts can be both summarize and observed in figure 9, by way of the invariant mass distributions. The blue histogram corresponds to all the events that have two clusters in the CsI and a charged particle in the gap. The grey, teal, violet, and dark colored histograms reflect the endpoint, M^2, and combined PID, cut respectively.

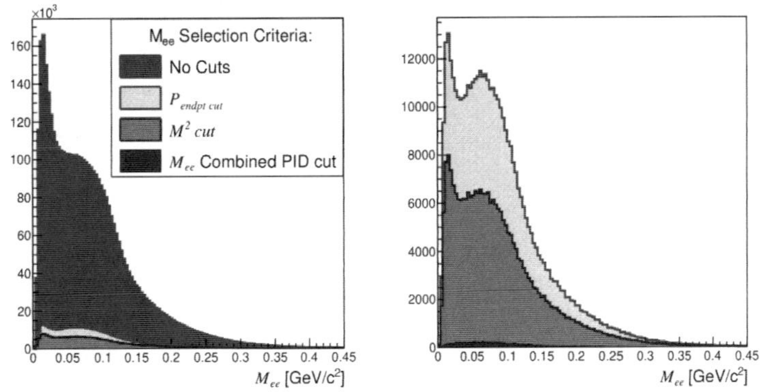

Figure 9: Invariant mass M_{ee} distributions under three cut conditions: good gap (blue); $K_{\pi 2}$ momentum region (green) and extra TOF1 (red).

The combined PID cut employs all the preceding cuts in addition to strict angular correlation between fired TOF1 counters and the corresponding CsI clusters. A fit to the M_{ee} for cuts of endpoint momentum cut in conjunction with M^2 cut, was performed and the results are show below in figure 10, with the residual, plotted in the panel below.

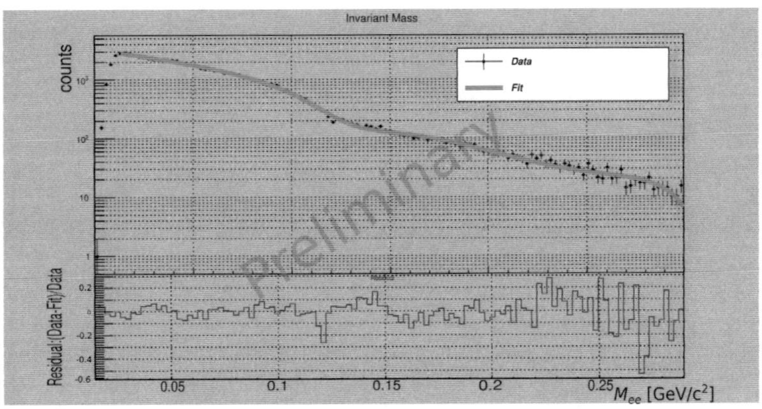

Figure 10: Invariant mass distribution with fit to data and residual is shown in the bottom plot.

The fit function from figure 10 and the total number of stopped kaons N_K were used to extract the upper limits. N_K, the total number of stopped K^+ in the fiducial region of the active target was computed as follows

$$N_K = \frac{N_{\mu 2}}{\mathcal{B}r(\mu 2) PS_{\mu 2} A_{\mu 2} LT_{\mu 2}} = 2.81 \times 10^9 \quad (5)$$

where $N_{\mu 2}$ is the number of tracked data candidates satisfying the $K_{\mu 2}$ decay momentum, $A_{\mu 2}$ is the acceptance of the $K_{\mu 2}$ muons evaluated with the MC simulations, $\mathcal{B}r(\mu 2)$ is the nominal branching ratio of the $K_{\mu 2}$ decay mode and $PS_{\mu 2} = 49$ and $LT_{\mu 2} = 1$ are the muon prescale factor and livetime fraction respectively. Upper limits at 95% CL on the branching ratio of $Br(A')$ for each A' mass, corresponding to a 2σ limit of not observing the A', under the assumption that $Br(A' \to e^+e^-) = 1$ were computed using the relation

$$Br(A') < \frac{2\sqrt{N_{\mu\nu ee}}}{N_K A_{A'} \cdot LT}. \tag{6}$$

The invariant mass M_{ee} spectrum reflecting the background reduction as result of the endpoint momentum cut, M^2 and TOF1 ≥ 3 is shown in figure 11 (a). The with $\sim$ 20% of data analysed, the upper limit was found to be at $\mathcal{O}(-6)$.

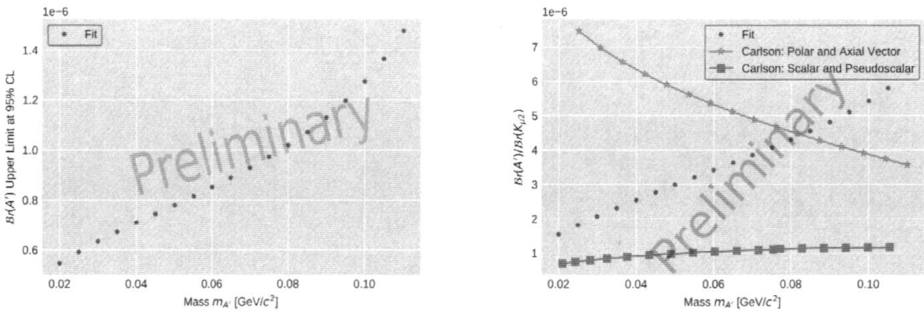

(a) (b)

Figure 11: Upper limit extraction (a) as a function of reconstructed A' mass. Data (blue) and theory mass limit comparisons (b) on particles with scalar and pseudoscalar (squares), and polar and axial vector (stars) couplings. Particles with both polar and axial vector coupling are excluded for masses below 0.08 GeV/c^2.

Experimental limits from the Carlson model for new light neutral bosons with polar (scalar) and axial vector (pseudoscalar) couplings are $\mathcal{O}(-6)$, see figure 2. Whereas figure 11 (b), shows the comparison between data and theory prediction of the mass limits on particles with scalar and pseudoscalar (green), and polar and axial vector (orange) couplings due to constraints placed by $K^+ \to \mu X$ searches [7]. A particle with both polar and axial vector coupling, is already excluded for masses below 0.08 GeV/c^2 with only $\sim$ 20% of the data analysed. However, a single particle with both scalar and pseudoscalar coupling is not yet excluded. With further development in reducible background suppression, and more data analysis, we are hoping to enhance sensitivity and enhance experimental reach. Presently data with a combined PID cut have not yet been assessed and the AC cuts have not been applied for $e/\mu/\pi$ mis-identification suppression. This cut would serve further reduce the reducible background, and would enhance our sensitivity to particles with scalar and pseudoscalar couplings.

5. Preliminary Results and Outlook

The TREK/E36 experiment has completed its data-taking and has been decomissioned as of March 2016. A Geant4 and ROOT based MC, e36g4MC, was developed to assist with rare kaon decay searches for light neutral bosons. A search for the A' production in the $K^+ \to \mu^+\nu A'$, followed by the prompt decay of $A' \to e^+e^-$ along with a charged particle that was detected and tracked in the spectrometer gap, was performed using data from the TREK/E36 experiment. The sensitivity of TREK/E36 to A' production was evaluated via upper limits of the branching ratios $Br(A')$ as a function of $m_{A'}$ and found were to be $\mathcal{O}(-6)$, with about 20% of the data analyzed. The experimental limit from the Carlson model for a new light neutral with polar (scalar) and axial vector (pseudoscalar) couplings are $\mathcal{O}(-6)$, see figure 2. Particles with both polar and axial vector coupling, is already excluded for masses below 0.08 GeV/c^2 with only $\sim$ 20% of the data analysed. However, a single particle with both scalar and pseudoscalar coupling is not yet excluded. Finally, a fully digitized and realistic e36g4MC would be instrumental in further studies of light neutral bosons. A fully digitized e36g4MC would be essential in providing a realistic description of detector responses as well as a full reproduction of known SM backgrounds. As a result the e36g4MC would play a pivotal role in reducing the SM backgrounds and increase our sensitivity to A' production.

6. References

[1] Vincenzo Cirigliano and Ignasi Rosell. Two-loop effective theory analysis of pi(K) → e anti-nu/e [gamma] branching ratios. *Phys. Rev. Lett.*, 99:231801, 2007.
[2] C. Rangacharyulu, et al. J-PARC Experimental Proposal Measurement of $\Gamma(K^+ \to e^+)/\Gamma(K^+ \to \mu^+\nu)$ and Search for heavy sterile neutrinos using the TREK detector system.
[3] Vincenzo Cirigliano and Ignasi Rosell. $\pi/K \to e\overline{\nu}_e$ branching ratios to $\mathcal{O}(e^2p^4)$ in Chiral Perturbation Theory. *JHEP*, 10:005, 2007.
[4] James D. Bjorken, Rouven Essig, Philip Schuster, and Natalia Toro. New Fixed-Target Experiments to Search for Dark Gauge Forces. *Phys. Rev.*, D80:075018, 2009.
[5] David Tucker-Smith and Itay Yavin. Muonic hydrogen and MeV forces. *Phys. Rev. D*, 83:101702, 2011.
[6] Brian Batell, David McKeen, and Maxim Pospelov. New Parity-Violating Muonic Forces and the Proton Charge Radius. *Phys. Rev. Lett.*, 107:011803, 2011.
[7] Carl E. Carlson and Benjamin C. Rislow. New Physics and the Proton Radius Problem. *Phys. Rev.*, D86:035013, 2012.
[8] Carl E. Carlson and Benjamin C. Rislow. Constraints to new physics models for the proton charge radius puzzle from the decay $K^+ \to \mu^+ + \nu + e^- + e^+$. *Phys. Rev. D*, 89(3):035003, 2014.
[9] Tommi Alanne, Giorgio Arcadi, Florian Goertz, Valentin Tenorth, and Stefan Vogl. Model-independent constraints with extended dark matter EFT. 6 2020.
[10] Marco Fabbrichesi, Emidio Gabrielli, and Gaia Lanfranchi. The Dark Photon. 5 2020.
[11] Carl E. Carlson and Michael Freid. Extending theories on muon-specific interactions. *Phys. Rev. D*, 92(9):095024, 2015.
[12] T. Beranek and M. Vanderhaeghen. Constraints on the Dark Photon Parameter Space from Leptonic Rare Kaon Decays. *Phys. Rev.*, D87(1):015024, 2013.
[13] K. Paton, et al. Measurement of T-violation Transverse Muon Polarization in $K^+ \to \pi^0\mu^+\nu$ Decays.
[14] O. Mineev et al. The design and basic performance of a Spiral Fiber Tracker for the J-PARC E36 experiment. *Nucl. Instrum. Meth.*, A847:136–141, 2017.
[15] Youichi Igarashi and Masatoshi Saito. Waveform sampler module for j-parc trek experiment. In *Waveform sampler module for J-PARC TREK experiment*, pages 1376–1379, 2012.
[16] D. V. Dementev et al. CsI(Tl) photon detector with PIN photodiode readout for a $K_{\mu 3}$ T-violation experiment. *Nucl. Instrum. Meth.*, A440:151–171, 2000.
[17] Y. Miyazaki et al. Performance test of a lead-glass counter for the J-PARC E36 experiment. *Nucl. Instrum. Meth.*, A779:13–17, 2015.

7. Acknowledgements

This work is supported in part by NSERC and NRC in Canada, the KAKEN-HI Grants-in-Aid for Scientific Research in Japan, the DOE and NSF in the US, and in Russia by the Ministry of Science and Technology.
Prepared by LLNL under Contract DE-AC52–07NA23744.

Sub-GeV dark matter at accelerators

Adam Ritz
Department of Physics and Astronomy, University of Victoria, Victoria, BC V8P 5C2, Canada
E-mail: aritz@uvic.ca

Abstract. Motivations for studying light MeV-to-GeV mass dark matter scenarios and opportunities for probing these models with high intensity fixed target experiments are reviewed. Attention is focussed on the lower $\mathcal{O}$(MeV) end of the thermal relic mass window, as relevant for high-intensity accelerators probing nuclear structure, and the complexities of models in this mass range due to cosmological constraints such as the bound on $N_{\rm eff}$. Existing limits, including those from direct detection electron-scattering experiments sensitive to a sub-leading but more energetic solar reflected component of the dark matter halo distribution, are also reviewed.

1. Introduction

Evidence for dark matter, through its gravitational impact in cosmology and astrophysics, remains one of the primary motivations for physics beyond the Standard Model. Among the most promising theoretical scenarios, the WIMP paradigm emerged from the observation that a weakly interacting thermal relic - with an abundance determined by freeze-out in the early universe - naturally forms a cold dark matter candidate. Given that the known relic energy density fixes the WIMP annihilation rate in the early universe, $\langle \sigma v\rangle \propto g^4 m_{\rm DM}^2/m_{\rm EW}^4 \sim \mathcal{O}(1\,{\rm pb})$, viability of the scenario rests on the dark matter mass $m_{\rm DM}$ being above the Lee-Weinberg bound of a few GeV. WIMPs remain a compelling dark matter scenario, given this minimality. However, it is now well appreciated that the paradigm of a thermal relic species freezing out of thermal equilibrium with the Standard Model bath is more general [1, 2, 3], provided that there are additional forces beyond the weak interactions, so-called ‘dark forces’. Importantly, this more general framework retains the insensitivity to early cosmology and initial conditions that a period of thermalization provides. It leads in turn to the idea of a multi-component ‘dark sector’, empirically motivated by both dark matter and neutrino mass, and there has been significant work exploring such hidden sectors experimentally over the past decade (see e.g. [4, 5, 10, 12]).

This generalization of the WIMP framework is still quite predictive, given a parametrization of the interactions between the SM and a hidden sector that assumes hidden sector states are SM gauge singlets. From the effective field theory expansion describing the interactions of light gauge singlet hidden sector fields with the SM,

$$\mathcal{L}_{\rm mediators} \sim \sum_{n=k+l-4} \frac{c_n}{\Lambda^n} \mathcal{O}^{(k)}_{\rm SM} \mathcal{O}^{(l)}_{\rm hidden}, \tag{1}$$

where $\mathcal{O}^{(k)}$ denote operators of dimension k, it follows that the lowest dimension interactions, namely those that are unsuppressed (other than logarithmically) by the heavy scale Λ, are preferentially probed at lower energy. The set of relevant or marginal interactions, usually

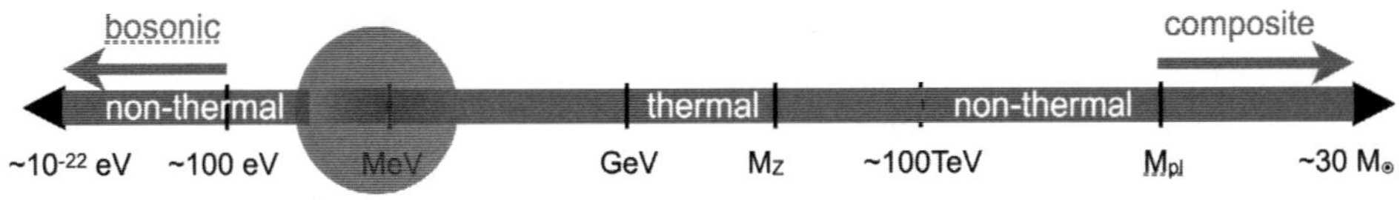

Figure 1. A schematic representation of the mass range for dark matter, with color used to reflect the production mode in the early universe. The mass window that is viable for thermal relics is shown in red, which expands the minimal WIMP window down to a low mass limit (circled) at the $\mathcal{O}$(MeV) scale, as dictated by constraints from BBN and the CMB (see Section 3 for a discussion).

termed 'portals', is quite compact. Up to dimension four ($n \leq 0$), assuming SM electroweak symmetry breaking, the list of portals is as follows [6, 7, 8, 9]:

Vector portal	$-\frac{\epsilon}{2}B_{\mu\nu}F'^{\mu\nu}$
Higgs portal	$(AS + \lambda S^2)H^\dagger H$
Neutrino portal	$y_N LHN$

where S, A'_μ and N are new 'dark' (SM-singlet) degrees of freedom coupled to SM operators involving the Higgs doublet H, hypercharge field strength $B_{\mu\nu}$, and the LH singlet. On general grounds, the coupling constants for these interactions are unsuppressed by any heavy scale of new physics, and thus it would be natural for new weakly-coupled physics to first manifest itself via these portals. Indeed, we observe that the right-handed neutrino coupling is amongst this list, which provides the simplest renormalizable model for neutrino mass and oscillations. It is natural to ask if the other portals are also realized, and in recent years all have been extensively studied in the dark matter context (see e.g. [4, 5, 10, 11, 12]).

This framework is theoretically quite compact, and it has driven a number of new experimental developments over the past decade. Conventional direct detection strategies utilizing elastic nuclear scattering are challenging for DM masses below a GeV, due to recoil energy thresholds and the reduced kinetic energy of the particles in the halo. This has motivated consideration of novel detection signatures, focussing on electron scattering, and also an expanded focus on high intensity accelerators and fixed target experiments, that can produce and detect light dark matter with relativistic kinematics, closer to that observed during early universe decoupling. In the remainder of this contribution, we review the status of accelerator–based probes of light sub–GeV dark matter, focussing on the low mass $\mathcal{O}$(MeV) boundary for a thermal relic (see Fig. 1), as dictated by limits from early cosmology. This overview will necessarily be schematic, particularly as regards the experiments providing the existing limits, and those proposed for future developments, and the reader is referred to the references for further details. We also discuss constraints from direct detection in this transitional mass range, which arise from sensitivity to electron scattering of a more energetic sub–component of the DM halo distribution produced through Solar rescattering.

2. Accelerator probes of light dark matter

Light thermal relic dark matter, with a mass below a few GeV, generically requires new annihilation channels with light mediators in order to reduce the relic abundance (mass density) down to the observed level. The simplest mediators couple via the renormalizable portal interactions, and the model that has become the standard benchmark, as it is the least constrained, utilizes a vector portal mediator A'_μ with the dark matter candidate a hidden scalar or fermion χ charged under $U(1)'$ [1, 2, 4, 5, 10, 11, 12]. At low energies, the Lagrangian is given by

$$\mathcal{L} = \mathcal{L}_\chi - \frac{1}{4}F'_{\mu\nu}F'^{\mu\nu} + \frac{1}{2}m^2_{A'}A'_\mu A'^\mu - \frac{\epsilon}{2}A'^{\mu\nu}F_{\mu\nu} + \cdots \tag{2}$$

with

$$\mathcal{L}_\chi = \begin{cases} i\bar{\chi}\not{D}\chi - m_\chi\bar{\chi}\chi, & \text{(Dirac fermion DM)} \\ |D_\mu\chi|^2 - m^2_\chi|\chi|^2, & \text{(Complex scalar DM)}, \end{cases}$$

where $D_\mu = \partial_\mu - ig_D A'_\mu$, with g_D the $U(1)'$ gauge coupling, and the ellipsis denotes terms associated with the spontaneous breaking of $U(1)'$, which will not be important here.

The vector mediator A'_μ may be on– or off–shell during production and scattering, but it is important to note that when kinematics allow it to decay on–shell to dark matter, $m_{A'} > 2m_\chi$, a light complex scalar DM candidate is less constrained by the impact of late annihilation on the CMB, as it is p–wave suppressed. Other variations of this model that can evade CMB constraints with fermionic DM include a particle–antiparticle asymmetry and/or psuedo–Dirac split states in the DM sector [13]. We also comment that the unbroken $U(1)'$, or $m_{A'} \to 0$, limit leads to a model of millicharged particles.

2.1. Accelerator production and detection modes

Having in mind a relatively low beam energy $E_{\rm beam} < \mathcal{O}(\rm GeV)$, as relevant for nuclear facilities, a number of DM production modes may be relevant:

- pseudoscalar meson decay, e.g. $\pi^0/\eta \to \gamma A'^* \to \gamma\chi\bar{\chi}$
- bremsstrahlung, e.g. $eN \to eNA'^* \to eN\chi\bar{\chi}$

For higher beam energies, Drell–Yan production at parton level also becomes important. Once produced the DM beam may be detected via a number of channels, depending on the detector:

- missing energy/mass/momentum
- neutral current–like scattering

The parameter space even for the minimal model is 4–dimensional $\{m_{A'}, m_\chi, \epsilon, \alpha_D\}$, and it has become standard to compare the reach of accelerator experiments in a 2–dimensional subspace of coupling versus DM mass. A convenient coupling to use is the canonical parameter $Y \equiv \epsilon^2\alpha_D(m_\chi/m_{A'})^4$, where $\alpha_D = g^2_D/(4\pi)$. This is motivated by considerations of DM production in the early universe. For $m_{A'} \gtrsim \mathcal{O}(\text{few}) \times m_\chi$, the DM annihilation rate near freeze–out is proportional to $\langle\sigma v\rangle \propto \alpha_{\rm em}\, Y/m^2_\chi$, implying that the cosmological χ abundance is in agreement with the observed DM energy density for

$$Y \sim \frac{m^2_\chi}{\alpha_{\rm em}\, T_{\rm eq}\, m_{\rm pl}} \sim \mathcal{O}(10^{-10}) \times \left(\frac{m_\chi}{100\ \text{MeV}}\right)^2, \tag{3}$$

where $T_{\rm eq}$ is the temperature at matter–radiation equality and $m_{\rm pl}$ is the Planck mass. This defines a *thermal target* in parameter space for accelerator–based and direct detection experiments (see Fig. 2 and 3).

Plotting Y against DM mass requires a choice of the ratio $R \equiv m_{A'}/m_\chi$, which controls whether production involves an on– or off–shell mediator, and also dictates the physics of the model in the early universe. We studied the sensitivity of a variety of experiments as a function of R, as shown in Fig. 2 [14], while conventional plots of Y vs m_χ have chosen the value $R = 3$, as shown in Fig. 3 [12]. Fig. 2 indicates that this choice is characteristic of the viable parameter regime. The current sensitivity limits are primarily determined at low mass by an NA64 missing mass search at CERN, older limits inferred from LSND electron scattering data, and by a BaBar monophoton analysis at higher mass.

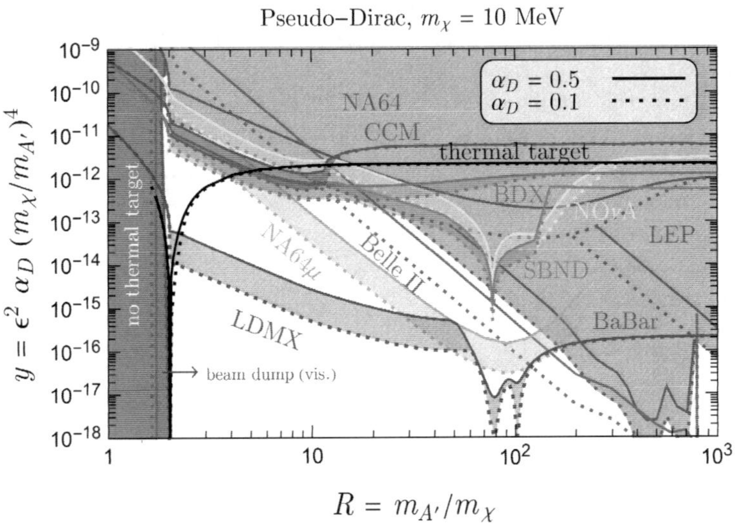

Figure 2. Constraints and future sensitivity projections in the DM parameter space as a function of Y and $R = m_{A'}/m_\chi$ for $m_\chi = 10$ MeV, from [14] (as updated in [11]).

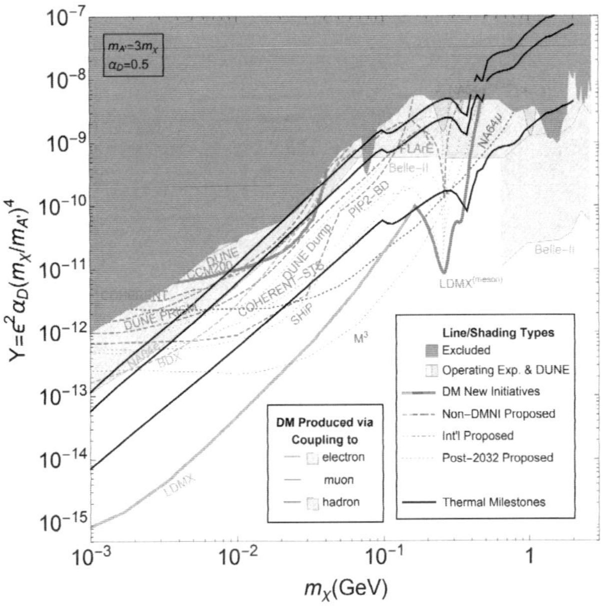

Figure 3. Constraints and future sensitivity projections in the DM parameter space as a function of Y and m_χ for $R = m_{A'}/m_\chi = 3$ from the recent Snowmass report [12].

3. The low mass $\mathcal{O}$(MeV) end of the thermal relic DM window

High–intensity accelerators operating at lower energies appropriate for the analysis of nuclear structure, such as the 520 MeV proton cyclotron and the 30–50 MeV e-linac at TRIUMF which are the focus of this workshop, motivate further consideration of DM models at the lower mass end of the thermal relic DM window.

Unlike the upper boundary of the thermal relic DM window, which is roughly defined by unitarity, the lower boundary is dictated primarily by stringent constraints from cosmology. The freezeout of a nonrelativistic relic, as required for structure formation, generally occurs at a temperature about an order of magnitude below its mass. As a consequence, thermal relics with a mass below about 10 MeV freezeout after the start of the critical epoch where in the standard cosmological model neutrino decoupling (at $T \sim 1 - 3$ MeV) triggers Big Bang nucleosynthesis (BBN). In particular, proton–neutron decoupling occurs at temperatures just below 1 MeV, with the deuterium bottleneck ending at $T \sim 100$ keV with the formation of the light elements. Energy injection into the SM plasma during (and following) this epoch is tightly constrained by the success of BBN and notably by the increasingly tight constraints from the CMB on additional radiation via $N_{\rm eff}$. The most recent analysis [15] of the coupled DM - neutrino - EM bath system leads to a bound $m_\chi > 7$ MeV for generic thermal relics. This can be relaxed for specific scenarios, e.g. sub–MeV thermal relic DM models remain feasible if the DM couples only to neutrinos [16] with late thermalization, but generally these constraints imply that DM below a mass of a few MeV requires non–thermal production mechanisms (as shown schematically in Fig. 1).

At the current time, the strongest accelerator–based constraints on models in this low mass range arise from analysis of electron scattering data at LSND, and more recent missing mass searches at NA64 (see Figs. 2 and 3). Future improvements in sensitivity appear possible at NA64e, COHERENT, CCM, SBND and DUNE at Fermilab, and LDMX and BDX if these projects proceed as planned [11, 12]. The best projected sensitivity at low mass is for LDMX at SLAC, and this project has received initial funding from DOE through the DMNI process. However, its reach relies on the success of a novel missing momentum technique.

4. Direct detection sensitivity to $\mathcal{O}$(MeV) mass DM

It is also useful to consider the status of direct detection sensitivity at the lower end of the thermal relic mass window. For DM below the electron mass threshold, it becomes more difficult to build thermal relic scenarios. However, very weakly coupled dark matter candidates can utilize non–thermal production, e.g. via gravitational production or the freeze–in mechanism, which generates the relic abundance via sub–thermal scattering off the SM bath. Direct detection experiments, focussed on electron recoil, have sensitivity to DM candidates across the lower MeV–scale boundary of the thermal relic DM mass range, due to the existence of sub–leading but more energetic components of the halo distribution, as discussed below. This sensitivity provides important complementarity to accelerator probes of DM in this $\mathcal{O}$(MeV) mass range.

4.1. Direct detection via Solar reflection and electron scattering

Direct detection experiments generally lose sensitivity rapidly once the mass of dark matter in the galactic halo falls below detector thresholds for energy deposited in scattering ($E_{\rm halo} \sim 10^{-6} m_{\rm DM}$). For nuclear scattering, this mass threshold is generally a few GeV, while for electron scattering the threshold is normally a few MeV [19]. Dark matter candidates χ, e.g. interacting via a dark vector, can have a mass below this threshold, and escape other direct or indirect constraints. However, even for dark matter below this mass threshold, as discussed in [17, 18] (see also [20]) a more energetic sub–component of the halo DM flux induced by re–scattering off thermal electrons in the Sun can be detectable. This 'reflected' component of the DM flux

Journal of Physics: Conference Series **2391** (2022) 012013 doi:10.1088/1742-6596/2391/1/012013

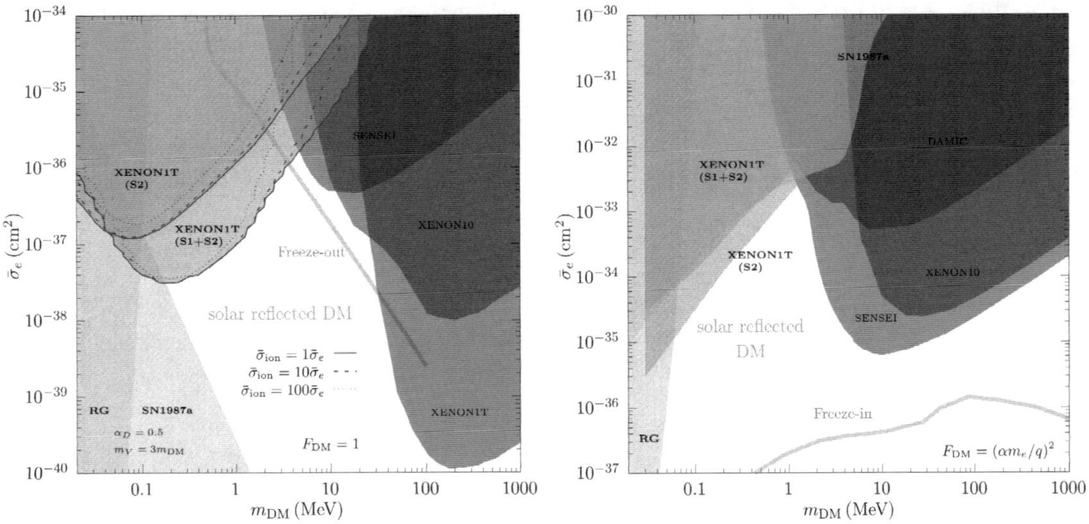

Figure 4. The limits on dark matter–electron scattering from solar reflection [17, 18]. The left–hand plot shows the limits in the heavy mediator limit, appropriate for a thermal relic, while the figure on the right shows the limit for DM in the freeze–in regime, with an ultralight mediator, leading to an enhanced form–factor for scattering at low velocities.

Φ can be approximated as follows,

$$\Phi_{\rm refl} \sim \frac{\Phi^{\rm halo}}{4} \times \begin{cases} \frac{4S_g}{3}\left(\frac{R_{\rm core}}{1\,{\rm A.U.}}\right)^2 \sigma_e n_e^{\rm core} R_{\rm core}, & \sigma_e \ll 1\,{\rm pb}, \\ S_g\left(\frac{R_{\rm scatt}}{1\,{\rm A.U.}}\right)^2, & \sigma_e \gg 1\,{\rm pb}, \end{cases} \tag{4}$$

where σ_e is the scattering cross section on electrons and $S_g \sim 10$ is a gravitational focussing factor. The energy of the reflected component satisfies $E_{\rm DM}^{\rm refl} < E_{\rm DM}^{\rm refl,max} = \frac{4E_e m_{\rm DM} m_e}{(m_e+m_{\rm DM})^2}$. This energy is generally above the ionization threshold for liquid xenon, and adds new sensitivity for any experiment able to detect electron ionization. The sensitivity of several experiments is shown in the left hand plot of Fig. 4, and can be improved in the near future.

The cross section evaluated at the atomic reference momentum transfer $q_0 = \alpha m_e$ for the benchmark model takes the form

$$\bar{\sigma}_e = \frac{16\pi\epsilon^2\alpha\alpha_D\mu_e^2}{(q_0^2+m_{A'}^2)^2} \sim \frac{16\pi\alpha Y m_e^2}{m_\chi^4} \sim 0.4 \times \left(\frac{Y}{10^{-10}}\right) \times \left(\frac{10\,{\rm MeV}}{m_\chi}\right)^4 \times 10^{-36}\,{\rm cm}^2, \tag{5}$$

where on the right hand side, we have assumed a heavy mediator so that $q_0 = \alpha m_e \ll m_{A'}$ and $m_e \ll m_\chi$. It follows from the left hand plot of Fig. 4 that the sensitivity at $m_\chi \sim 10$ MeV is at the level of $Y \sim 10^{-10}$, somewhat below but comparable to the sensitivity exhibited by accelerators from Fig. 3. The differing assumptions limit a more general comparison, but this commensurate sensitivity emphasizes the complementarity of both detection strategies.

The scenario above assumes contact interactions between dark matter and electrons. However, while natural for thermal relic candidates, this assumption is less appropriate for models with mass below the thermal relic window, which rely on non–thermal production mechanisms, e.g. freeze–in. This mechanism assumes that the DM relic density is built up through weak scattering off the SM bath, and thermal equilibrium is never established. The benchmark vector portal DM model can be extended to sub–MeV masses in this way, but the couplings are much smaller. Thus sensitivity in terrestrial experiments is reduced unless the mediator is parametrically light, where low velocity scattering is enhanced due to the effective long–range interaction. This regime requires a dedicated analysis in computing the Solar reflected flux, and the ensuing sensitivity is highlighted in the right hand plot of Fig. 4 [18]. For such non–thermal models, direct detection has substantial advantages over accelerator probes, as the enhancement in low velocity scattering is not present with a relativistic beam.

Acknowledgments

I'd like to thank the organizers for their invitation to this workshop, and my co–authors, H. An, A. Berlin, P. deNiverville, H. Nie, M. Pospelov, J. Pradler, P. Schuster, and N. Toro for fruitful and enjoyable collaborations on the work described here, which was supported in part by NSERC, Canada.

References

[1] Boehm C and Fayet P 2004 *Nucl. Phys.* **B683** 219–263 (*Preprint* hep-ph/0305261)
[2] Pospelov M, Ritz A and Voloshin M B 2008 *Phys. Lett.* **B662** 53–61 (*Preprint* 0711.4866)
[3] Feng J L and Kumar J 2008 *Phys. Rev. Lett.* **101** 231301 (*Preprint* 0803.4196)
[4] Alexander J *et al.* 2016 Dark Sectors 2016 Workshop: Community Report (*Preprint* 1608.08632)
[5] Battaglieri M *et al.* 2017 (*Preprint* 1707.04591)
[6] Holdom B 1986 *Phys. Lett.* **166B** 196–198
[7] Patt B and Wilczek F 2006 (*Preprint* hep-ph/0605188)
[8] Foot R, Lew H and Volkas R R 1991 *Phys. Lett.* **B272** 67–70
[9] Batell B, Pospelov M and Ritz A 2009 *Phys. Rev.* **D80** 095024 (*Preprint* 0906.5614)
[10] Beacham J *et al.* 2020 *J. Phys. G* **47** 010501 (*Preprint* 1901.09966)
[11] Krnjaic G *et al.* 2022 (*Preprint* 2207.00597)
[12] Gori S *et al.* 2022 (*Preprint* 2209.04671)
[13] Izaguirre E, Krnjaic G, Schuster P and Toro N 2014 *Phys. Rev. D* **90** 014052 (*Preprint* 1403.6826)
[14] Berlin A, deNiverville P, Ritz A, Schuster P and Toro N 2020 *Phys. Rev. D* **102** 095011 (*Preprint* 2003.03379)
[15] Chu X, Kuo J L and Pradler J 2022 *Phys. Rev. D* **106** 055022 (*Preprint* 2205.05714)
[16] Berlin A and Blinov N 2018 *Phys. Rev. Lett.* **120** 021801 (*Preprint* 1706.07046)
[17] An H, Pospelov M, Pradler J and Ritz A 2018 *Phys. Rev. Lett.* **120** 141801 (*Preprint* 1708.03642)
[18] An H, Nie H, Pospelov M, Pradler J and Ritz A 2021 *Phys. Rev. D* **104** 103026 (*Preprint* 2108.10332)
[19] Essig R, Manalaysay A, Mardon J, Sorensen P and Volansky T 2012 *Phys. Rev. Lett.* **109** 021301 (*Preprint* 1206.2644)
[20] Emken T, Kouvaris C and Nielsen N G 2018 *Phys. Rev.* **D97** 063007 (*Preprint* 1709.06573)

Probing Lepton Flavor Universality with Pion Decays

Douglas Bryman

Department of Physics and Astronomy, University of British Columbia, Vancouver, BC, Canada V6T 1Z1

TRIUMF, Vancouver, BC, Canada V6T 2A3

E-mail: doug@triumf.ca

Abstract. Searches for violation of lepton flavor universality are topical due to hints of discrepancies with Standard Model expectations in studies involving muons, B mesons, and beta decays which may be related by tests of the unitarity of the Cabibbo-Kobayashi-Maskawa quark mixing matrix. The status and prospects of pion decay experiments which offer the most precise tests of $e - \mu$ universality are discussed.

1. Introduction

Since the Standard Model (SM) is clearly not the final word on the fundamental theory, the search for new physics effects proceeds on many fronts. New particles and interactions are being sought directly at high energy colliders and indirectly in searches using numerous techniques (see reviews in Ref. [1]). Indirect searches for violations of symmetries are among the approaches being pursued due to the potential sensitivity to high mass scales. One approximate symmetry built into the SM is lepton flavor universality (LFU). Violation of LFU (LFUV) could occur, for example, if there were interactions with different weak couplings for electrons, muons and tau leptons. LFUV can be probed through a wide range of measurements (see [2] and references therein). Experiments dealing with meson, tau, and beta decays are being pursued vigorously in light of hints of possible new flavor physics effects and the possibly related deviation in first-row unitarity of the Cabibbo-Kobayashi-Maskawa (CKM) matrix [3, 4]. In this paper, the sensitivity of current and proposed rare pion decay measurements to LFUV and other non-SM effects will be briefly outlined.

2. Rare Pion Decays and Lepton Universality

As discussed in Ref. [2], the helicity suppressed ratio of pion decay rates

$$R^{\pi}_{e/\mu} = \Gamma[\pi \to e\bar{\nu}_e(\gamma)]/\Gamma[\pi \to \mu\bar{\nu}_\mu(\gamma)]\,, \tag{1}$$

provides the most stringent test of LFU because it can be calculated with extraordinary precision at the 10^{-4} level and can also be measured accurately. $R^{\pi}_{e/\mu}$ is a sensitive probe of non-universal corrections to W-lepton couplings especially if they generate a pseudo-scalar current or an induced scalar current [5]. The exceptionally accurate theoretical calculations of $R^{\pi}_{e/\mu}$ [6, 7, 8, 9] involve the application of chiral perturbation theory (ChPT) and include structure dependent contributions to $\pi \to \ell\bar{\nu}_\ell\gamma$ [10] which are not helicity suppressed. The SM expectation[2] is

$$R(\mathrm{SM})^{\pi}_{e/\mu} = 1.23524(15) \times 10^{-4} \tag{2}$$

where the dominant uncertainty arises from a low-energy constant in ChPT, followed by non-leading logarithmic corrections of $O(\alpha^2)$.

The most precise measurement of $R^{\pi}_{e/\mu}$ was reported by the TRIUMF PIENU experiment [11] which will be discussed below. The result, based on analysis of an initial data set, was

$$R(\mathrm{Exp})^{\pi}_{e/\mu} = (1.2344 \pm 0.0023(\mathrm{stat}) \pm 0.0019(\mathrm{syst})) \times 10^{-4}. \tag{3}$$

The PDG [1] average including previous experiments done at TRIUMF [11, 12, 13] and PSI [14] is

$$R(\overline{\mathrm{Exp}})^{\pi}_{e/\mu} = (1.2327 \pm 0.0023) \times 10^{-4}. \tag{4}$$

The comparison between theory and experiment given in Eqs. (2) and (4) provides the most precise measure of e–μ universality. The ratio of effective couplings A_ℓ ($\ell = e, \mu$) multiplying the low-energy charged current contact interaction obtained using $R^{\pi}_{e/\mu}$ is

$$\left(\frac{A_\mu}{A_e}\right)_{R^{\pi}_{e/\mu}} = 1.0010 \pm 0.0009\,, \tag{5}$$

in excellent agreement with the SM expectation. Deviation from $A_\ell/A_{\ell'} = 1$ could be due to various mechanisms such as flavor-dependent couplings of the W-boson to the leptonic current. In the context of modified W-boson couplings, LFU tested with $R^{\pi}_{e/\mu}$ probes the couplings of a longitudinally polarized W-boson whereas purely leptonic reactions like $\tau \to \ell\nu_\tau\nu_\ell$ give comparable precision[2] on the couplings of the transversely polarized W-boson. Apparent deviations from $A_\ell/A_{\ell'} = 1$ could also be due the presence of unaccounted-for sterile neutrinos[15].

Charged μ-e universality can also be tested using similar K and W decays[2] and B decays such as $\mathrm{Br}[B \to D^*\mu\nu]/\mathrm{Br}[B \to D^*e\nu]$ [16, 17, 18]; these are interesting in light of possible anomalies in $R(D^{(*)})$=$\mathrm{Br}[B \to D^{(*)}\tau\nu_\tau]/\mathrm{Br}[B \to D^{(*)}\ell\nu_\ell]$ [19, 20, 21] , and $R(K^{(*)})$=$\mathrm{Br}[B \to K^{(*)}\mu^+\mu^-]/\mathrm{Br}[B \to K^{(*)}e^+e^-]$ [22, 23, 24] which reflect the impact of possible four-fermion operators involving leptons and heavy quarks. CKM unitarity tests which rely principally on beta decays and K decays for inputs to the extraction of V_{ud} and V_{us} may be related to LFU since calculation of V_{ud} relies on obtaining G_F from μ decay (see Refs. [4, 2] and references therein). Recently it has been pointed out [25] that V_{us}/V_{ud} can be obtained using the ratio of decay rates $R_V \equiv \Gamma(K \to \pi\ell\nu(\gamma))/\Gamma(\pi^+ \to \pi^0e^+\nu(\gamma))$ where the uncertainty is dominated by pion beta decay which is discussed below. As illustrated in Fig. 1[2] tensions exist among current determinations of V_{us} and with overall CKM unitarity.

 IOP Publishing
2391 (2022) 012014 doi:10.1088/1742-6596/2391/1/012014

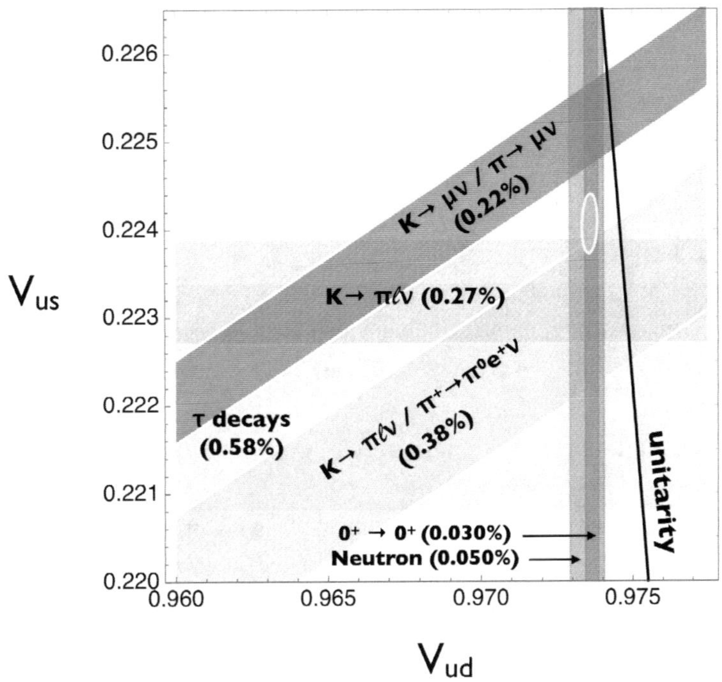

Figure 1. Summary of constraints on V_{ud} and V_{us} (assuming the Standard Model hypothesis) from nuclear, nucleon, meson, and τ lepton decays. For each constraint, the one-sigma uncertainty on V_{us} or V_{ud} is given in parenthesis (see [2] for details). The one-sigma ellipse from a global fit (with χ^2/d.o.f. = 2.8), depicted in yellow, corresponds to $V_{ud} = 0.97357(27)$ and $V_{us} = 0.22406(34)$, implying $\Delta_{\text{CKM}} = |V_{ud}|^2 + |V_{us}|^2 - 1 = (-19.5 \pm 5.3) \times 10^{-4}$. Figure reproduced from [2].

3. Rare Pion Decay Experiments

The TRIUMF PIENU experiment [11, 26] used pions stopped in plastic scintillator and measured decay positrons in a calorimeter consisting of NaI(Tl) and pure CsI crystals. The branching ratio given in Eq. 3 was obtained by separating events into high- and low-energy regions corresponding to the positron energy spectra of $\pi^+ \to e^+\nu$ and $\pi^+ \to \mu^+ \to e^+$ decays. The time spectra were fit in each region with the $\pi^+ \to e^+\nu$ and $\pi^+ \to \mu^+ \to e^+$ decay shapes as well as backgrounds originating from sources including pion decays in flight and pile-up contamination from decays of residual muons. The raw branching ratio $R^{raw}_{e/\mu}$ was the ratio of the $\pi^+ \to e^+\nu$ amplitude to the $\pi^+ \to \mu^+ \to e^+$ amplitude. Corrections such as the tail correction for low energy $\pi \to e\nu$ events below the $\mu \to e\bar{\nu}\nu$ spectrum, principally due to electromagnetic shower leakage, were subsequently applied to obtain the final value. PIENU and the PEN [27, 28, 29] experiment (which used a similar technique) aim to further improve the precision of the $\pi \to e\nu$ branching ratio $R^{\pi}_{e/\mu}$ using larger available data sets. However, even when these goals are realized, an experimental improvement by more than an order of magnitude in uncertainty would be valuable to confront the SM prediction (Eq. 2) and to search for non-SM effects.

The recently proposed PIONEER experiment[30] at the Paul Scherrer Institut (PSI) aims for an improvement in precision for $R^{\pi}_{e/\mu}$ by an order of magnitude making the experimental uncertainty comparable to the theoretical uncertainty in Eq. 2. To reach very high precision requires high statistics as well as reduction of systematic uncertainties, backgrounds, biases, and distortions in the data selection criteria. Like PIENU [11] and PEN [27], PIONEER will be done using stopped pions that decay at rest.

The PIONEER detector[30] is illustrated in Fig. 2. An intense pion beam is brought to rest in an instrumented target detector (ATAR) and an electromagnetic calorimeter (CALO) surrounds the stopping target. A cylindrical tracker surrounding the ATAR is used to link the locations of pions stopping in the target to showers in the calorimeter. Features of the PIONEER approach will include improved time and energy resolutions, increased calorimeter depth, high-speed detector and electronic response, large solid angle coverage, and complete event reconstruction. The proposed detector will include a 3π sr, 25 radiation length (X_0) LXe electromagnetic calorimeter, similar in concept to the MEG LXe calorimeter[31]; an alternate crystal-based calorimeter design is also under consideration. The 2×2 cm^2 ATAR is an advanced design fully active LGAD Si-strip detector[32] which will provide fine-grained position and energy-loss measurements for incoming pions, muons from pion decay, and outgoing positrons. The preliminary ATAR geometry has 48 planes, each with 100 $120\mu m$-thick strips on a pitch of $200\mu m$; planes will alternate between horizontal and vertical directions.

PIONEER aims to improve both the statistical and systematic uncertainties in the measurement of $R^{\pi}_{e/\mu}$. In particular, the ATAR particle identification and tracking capabilities will enable improved precision on the tail correction mentioned above, which was the largest source of systematic uncertainty in the PIENU measurement [11]. Figure 3(Left) shows the simulated detector response for $\mu \to e\nu\overline{\nu}$ and $\pi \to e\nu$ decays and the background-suppressed distributions (Right) used to determine the tail correction. The remaining small background is dominated by events with $\pi \to \mu$ decay at rest (DAR) followed by $\mu \to e$ decay decay in flight (DIF) before stopping; it is anticipated that this component can be independently measured. Table 1 gives the estimated uncertainties for the PIONEER measurement of $R^{\pi}_{e/\mu}$ in comparison with [11].

PIONEER also aims to improve the precision for the branching ratio of pion beta decay $\pi^+ \to \pi^0 e^+\nu(\gamma)$ [33] by a factor of three (ten) in the second (third) phase of the experiment. Pion beta decay, while providing the theoretically cleanest method to obtain $|V_{ud}|$, requires a 10x improvment in the experimental precision to be competitive. However, a 3x improvement in precision of the pion beta decay branching ratio compared to Ref. [33] would allow for a 0.2% accuracy in the determination of the ratio V_{us}/V_{ud}, competitive with the existing determination[25].

Searches for heavy sterile neutrinos ν_H in $\pi^+ \to \ell^+\nu_H$ decays were also performed in the PIENU experiment [34, 35]. PIONEER aims to make improvements in sensitivity by an order of magnitude for processes like $\pi^+ \to e^+\nu_H$ decay as indicated in Figure 4. Similarly, pion decays involving massive or massless weakly interacting neutral bosons X such as axions [36, 37, 38, 39]

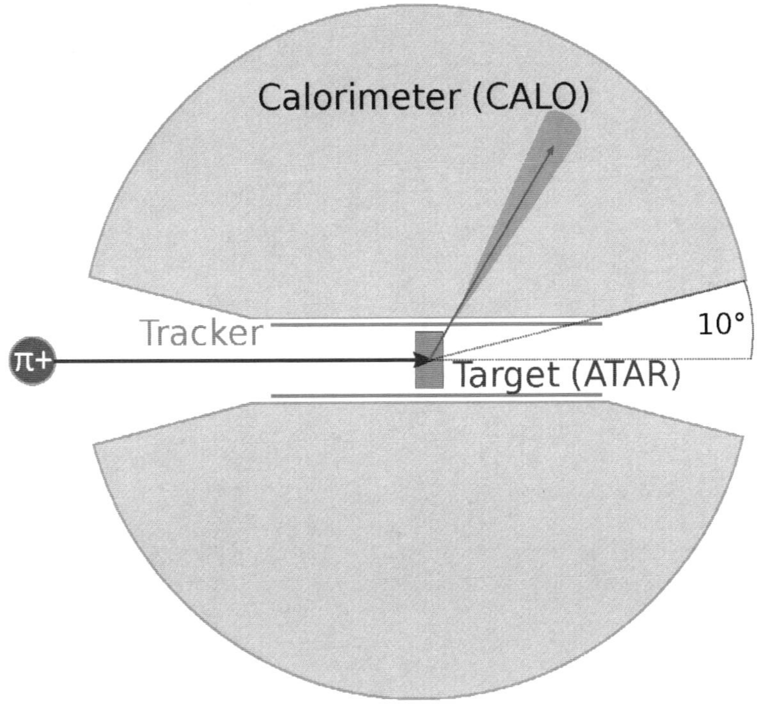

Figure 2. Conceptual layout of the PIONEER rare pion decay experiment[30]. The intense positive pion beam enters from the left and is brought to rest in a highly segmented active target (ATAR). Decay positron trajectories are measured from the ATAR to an outer electromagnetic calorimeter (CALO) through a tracker. The CALO records the energy, time, and location of positron tracks. An example outgoing positron track is shown in green.

Error Source	PIENU 2015 %	PIONEER Estimate %
Statistics	0.19	0.007
Tail Correction	0.12	<0.01
t_0 Correction	0.05	<0.01
Muon DIF	0.05	0.005
Parameter Fitting	0.05	<0.01
Selection Cuts	0.04	<0.01
Acceptance Correction	0.03	0.003
Total Uncertainty	**0.24**	$\leq$ **0.01**

Table 1. $R^{\pi}_{e/\mu}$ precision reported by PIENU[11] and estimated precision for PIONEER[30].

and Majorons [40, 41, 42] have been searched for; these may involve dark matter candidates, or impact baryogenesis or the strong CP problems [43]. Other reactions of interest include rare pion decays $\pi^+ \to \ell^+\nu_\ell\nu\bar{\nu}$ which are highly suppressed; the experimental search for these processes could reveal small non-SM effects such as neutrino-neutrino interactions [44] and six-fermion

Journal of Physics: Conference Series **2391** (2022) 012014 doi:10.1088/1742-6596/2391/1/012014

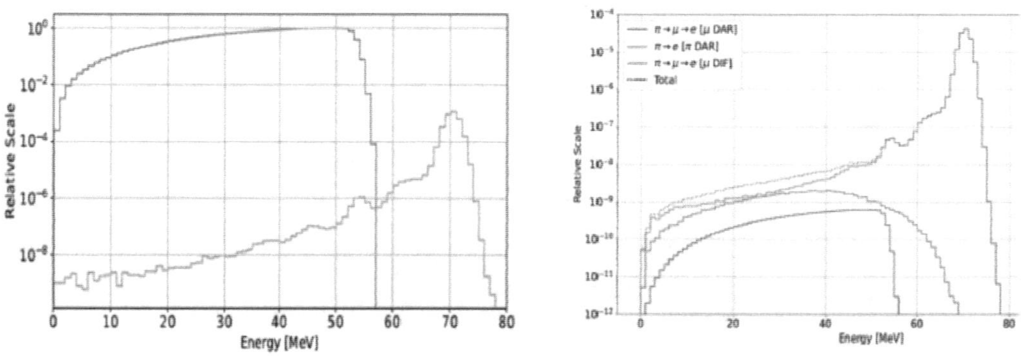

Figure 3. Left: Simulated decay spectra for $\mu \to e\nu\overline{\nu}$ (blue) and $\pi \to e\nu$ (orange) decays. Right: $\mu \to e$ background-suppressed spectra showing the components due to $\pi \to e\nu$ decays (orange) and two remaining backgrounds: $\mu \to e$ DIF after $\pi \to \mu$ DAR (green), and $\mu \to e$ DAR (blue) from $\pi \to \mu$ DIF[30]. See text.

interactions [45, 46], which might compete with the SM processes at first order; considering three models (SM, neutrino-neutrino interaction, and six-fermion) these latter processes were also searched for in PIENU [47] and sensitivity to them will be improved in PIONEER.

4. Conclusions

The status of the searches for violation of lepton flavor universality in the charged current involving pions decays has been discussed. This high precision test of LFU agrees well with the SM expectation at the $O(10^{-3})$ level and is particularly interesting in light of the experimental hints for possible large deviations from LFU in semi-leptonic B decays, the anomalous magnetic moment of the muon[49], and the Cabbibo angle anomaly. The proposed PIONEER experiment aims to improve the test of $e - \mu$ universality in pion decay, the pion beta decay determination of V_{ud}, and searches for exotic decays by an order of magnitude.

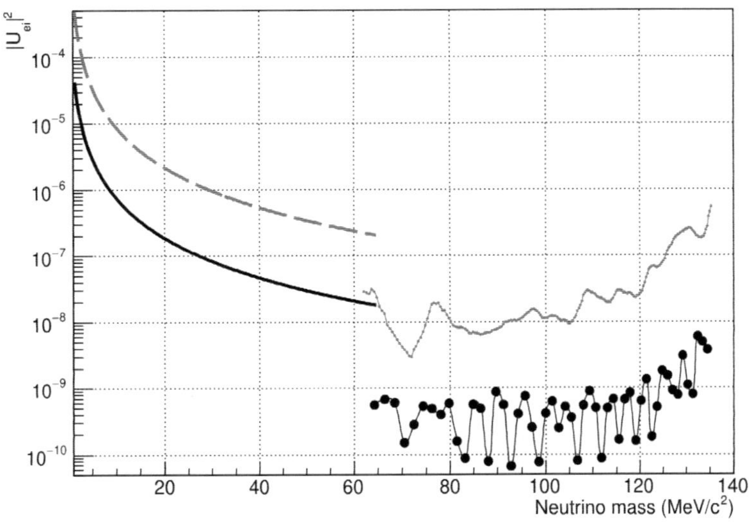

Figure 4. Limits on mixing coefficients for fourth generation neutrinos coupling to electrons $|U_{ei}|^2$ (90% C.L.) from PIENU (red)[11, 34]) and expected sensitivity from PIONEER (black). The lower mass region limits (m_{ν_H} <65 MeV) come from the $R^{\pi}_{e/\mu}$ branching ratio measurement and the higher mass region limits (m_{ν_H} >65 MeV) from a peak search. Figure reproduced from Ref. [48].

Acknowledgments
The author thanks the PIENU and PIONEER collaborations. This work is supported by NSERC grant no. SAPPJ-2018-0017 (Canada).

References
[1] Zyla P A *et al.* (Particle Data Group) 2020 *PTEP* **2020** 083C01
[2] Bryman D, Cirigliano V, Crivellin A and Inguglia G 2021 (*Preprint* 2111.05338)
[3] Coutinho A M, Crivellin A and Manzari C A 2020 *Phys. Rev. Lett.* **125** 071802 (*Preprint* 1912.08823)
[4] Crivellin A and Hoferichter M 2020 *Phys. Rev. Lett.* **125** 111801 (*Preprint* 2002.07184)
[5] Campbell B A and Maybury D W 2005 *Nucl. Phys. B* **709** 419–439 (*Preprint* hep-ph/0303046)
[6] Marciano W J and Sirlin A 1993 *Phys. Rev. Lett.* **71** 3629–3632
[7] Finkemeier M 1996 *Phys. Lett. B* **387** 391–394 (*Preprint* hep-ph/9505434)
[8] Cirigliano V and Rosell I 2007 *Phys. Rev. Lett.* **99** 231801 (*Preprint* 0707.3439)
[9] Cirigliano V and Rosell I 2007 *JHEP* **10** 005 (*Preprint* 0707.4464)
[10] Bijnens J, Ecker G and Gasser J 1993 *Nucl. Phys. B* **396** 81–118 (*Preprint* hep-ph/9209261)
[11] Aguilar-Arevalo A *et al.* (PIENU) 2015 *Phys. Rev. Lett.* **115** 071801 (*Preprint* 1506.05845)
[12] Bryman D A, Dixit M S, Dubois R, Macdonald J A, Numao T, Olaniyi B, Olin A and Poutissou J M 1986 *Phys. Rev. D* **33** 1211
[13] Britton D I *et al.* 1994 *Phys. Rev. D* **49** 28–39
[14] Czapek G *et al.* 1993 *Phys. Rev. Lett.* **70** 17–20
[15] Bryman D A and Shrock R 2019 *Phys. Rev. D* **100** 073011 (*Preprint* 1909.11198)
[16] Glattauer R *et al.* (Belle) 2016 *Phys. Rev. D* **93** 032006 (*Preprint* 1510.03657)

[17] Abdesselam A *et al.* (Belle) 2017 (*Preprint* 1702.01521)
[18] Waheed E *et al.* (Belle) 2019 *Phys. Rev. D* **100** 052007 [Erratum: Phys.Rev.D 103, 079901 (2021)] (*Preprint* 1809.03290)
[19] Lees J P *et al.* (BaBar) 2012 *Phys. Rev. Lett.* **109** 101802 (*Preprint* 1205.5442)
[20] Aaij R *et al.* (LHCb) 2018 *Phys. Rev. D* **97** 072013 (*Preprint* 1711.02505)
[21] Abdesselam A *et al.* (Belle) 2019 (*Preprint* 1904.08794)
[22] Aaij R *et al.* (LHCb) 2017 *JHEP* **08** 055 (*Preprint* 1705.05802)
[23] Aaij R *et al.* (LHCb) 2019 *Phys. Rev. Lett.* **122** 191801 (*Preprint* 1903.09252)
[24] Aaij R *et al.* (LHCb) 2021 (*Preprint* 2103.11769)
[25] Czarnecki A, Marciano W J and Sirlin A 2020 *Phys. Rev. D* **101** 091301 (*Preprint* 1911.04685)
[26] Aguilar-Arevalo A *et al.* 2010 *Nucl. Instrum. Meth. A* **621** 188–191 (*Preprint* 1003.2235)
[27] PEN Collaboration http://pen.phys.virginia.edu/
[28] Glaser C J *et al.* (PEN) 2018 PEN experiment: a precise test of lepton universality *13th Conference on the Intersections of Particle and Nuclear Physics* (*Preprint* 1812.00782)
[29] Pocanic D, Frlez E and van der Schaaf A 2014 *J. Phys. G* **41** 114002 (*Preprint* 1407.2865)
[30] Altmannshofer W *et al.* (PIONEER) 2022 (*Preprint* 2203.01981)
[31] Mihara S 2011 *J. Phys. Conf. Ser.* **308** 012009
[32] Mazza S M (PIONEER) 2021 *Instruments* **5** 40 (*Preprint* 2111.05375)
[33] Pocanic D *et al.* 2004 *Phys. Rev. Lett.* **93** 181803 (*Preprint* hep-ex/0312030)
[34] Aguilar-Arevalo A *et al.* (PIENU) 2018 *Phys. Rev. D* **97** 072012 (*Preprint* 1712.03275)
[35] Aguilar-Arevalo A *et al.* (PIENU) 2019 *Phys. Lett. B* **798** 134980 (*Preprint* 1904.03269)
[36] Jaeckel J and Ringwald A 2010 *Ann. Rev. Nucl. Part. Sci.* **60** 405–437 (*Preprint* 1002.0329)
[37] Jeong K S, Jung T H and Shin C S 2020 *Phys. Rev. D* **101** 035009 (*Preprint* 1811.03294)
[38] Agrawal P and Howe K 2018 *JHEP* **12** 029 (*Preprint* 1710.04213)
[39] Alves D S M and Weiner N 2018 *JHEP* **07** 092 (*Preprint* 1710.03764)
[40] Gelmini G B and Roncadelli M 1981 *Phys. Lett. B* **99** 411–415
[41] Chikashige Y, Mohapatra R N and Peccei R D 1981 *Phys. Lett. B* **98** 265–268
[42] Aulakh C S and Mohapatra R N 1982 *Phys. Lett. B* **119** 136–140
[43] Wilczek F 1982 *Phys. Rev. Lett.* **49** 1549–1552
[44] Bardin D Y, Bilenky S M and Pontecorvo B 1970 *Phys. Lett. B* **32** 121–124
[45] Ericson T E O and Glashow S L 1964 *Phys. Rev.* **133** B130–B131
[46] Vanzha A, Isaev A and Lapidus L 1971 *Sov. J. Nucl. Phys.* **12** 325 [Yad. Fiz. **12**, 595 (1970)]
[47] Aguilar-Arevalo A *et al.* (PIENU) 2020 *Phys. Rev. D* **102** 012001 (*Preprint* 2006.00389)
[48] Altmannshofer W *et al.* (PIONEER) 2022 Testing Lepton Flavor Universality and CKM Unitarity with Rare Pion Decays in the PIONEER experiment *2022 Snowmass Summer Study* (*Preprint* 2203.05505)
[49] Abi B *et al.* (Muon g-2) 2021 *Phys. Rev. Lett.* **126** 141801 (*Preprint* 2104.03281)

Lepton Universality Test with MUSE at PSI

Michael Kohl
for the MUSE Collaboration
Physics Department, Hampton University
E-mail: kohlm@jlab.org

Abstract. Lepton universality (LU) typically refers to the lepton coupling, which is considered to be the same for e, μ, and τ leptons, if the interaction is electroweak according to the Standard Model, and it is hence a compelling probe for New Physics. The same principle of universal electroweak lepton interaction leads to the expectation that lepton scattering yields are equal for e and μ beams under the same kinematic condition. The mere mass difference between e and μ affects kinematic quantities (such as the relation between scattering angle and Q^2), and the lepton mass dependence of elastic cross sections for leptons scattered from structured and pointlike objects are taken into account. By comparing e^+, e^-, μ^+, and μ^- scattering yields, two-photon exchange (TPE) effects, universal or not, can be separated from the general LU test of the e/μ yield ratio. With its separable mixed beams of e^+/μ^+ and e^-/μ^-, respectively, the MUSE experiment at PSI is not only designed to measure the proton charge radius with four lepton species, but is also uniquely suited to probe TPE and LU, while benefitting from partial cancellations of certain shared systematics. An overview will be given of the MUSE experiment, the sensitivity, and the present status.

1. Introduction

The Proton Radius Puzzle (PRP) began in 2010 with the then two most precise measurements of the proton charge radius [1, 2] contradicting each other at the 5-7 σ level [3, 4] or by 4% of the radius value. One measurement was based on a novel method using the Lamb shift in muonic hydrogen with enhanced sensitivity to the finite proton size effect on the 2S-2P transitions, resulting in a radius of about 0.84 fm, substantially smaller and more precise than the previous world average [1, 5]. The other measurement used a very large and comprehensive data set obtained with traditional electron scattering from hydrogen with high resolution spectrometers, resulting in a larger radius value of about 0.88 fm and consistent with the previous world average from decades of atomic hydrogen spectroscopy and electron scattering data [2, 6]. Several following experiments using scattering and spectroscopy methods, as well as a number of global analyses, obtained radius values covering the entire size of the puzzle, with the present situation remaining inconclusive [7]. If the discrepancy given by the two data sets from 2010 is true, it could hint to signals beyond the Standard Model in the muon-electron sector of lepton non-universality.

The MUSE experiment at Paul-Scherrer Institute (PSI) was conceived upon the fact that there have been precise radius determinations with spectroscopy both from regular and muonic hydrogen, while only electrons had been used for radius measurements from elastic scattering, and there has been no data from muon scattering with enough precision to discriminate between the contradicting data sets. Muon beams are necessarily generated as secondary beams with properties different from electron beams in electron accelerators. The secondary beam of the piM1 beamline at PSI includes besides muons also electrons and pions. A magnetic chicane is used to prepare the secondary beam with chosen charge polarity and selected momentum bite. The beam intensity is typically a few MHz with a 20 ns repetition rate for the primary proton beam in continuous wave mode. The e, π, and μ particles arriving at the MUSE experiment differ in their arrival time due to different average velocities of the three particle types. Three beam momenta of 115, 160, and 210 MeV/c are chosen for which the separation in arrival times between e, π, and μ is maximized to become several ns within the 20-ns repeating window. Secondary beams, like the one used at the piM1 beamline, are typically focussed to a finger-width beamspot and exhibit significant angular variation within the beam envelope, and thus require event-by-event beam particle tracking to provide the reference for an accurate scattering angle determination with the scattered particle detection system. The properties of the piM1 beamline at PSI used for the MUSE experiment have been fully characterized experimentally and matched with dedicated simulations [8].

2. MUSE setup

A schematic view of the MUSE setup is shown in Fig. 1, consisting of a mixed, collimated $e/\mu/\pi$ beam, scintillating-fiber beam hodoscopes (BH) for time-of-flight and position, Gas Electron Multiplier (GEM) tracking detectors for precise beam particle position and angle, a liquid hydrogen target, as well as drift tube chambers (STT) and scintillator walls (SPS) to detect scattered particles. Located downstream of the target is a beam monitor hodoscope (BM) for luminosity monitoring and quality control, and an electromagnetic calorimeter (CALO) to control radiative corrections.

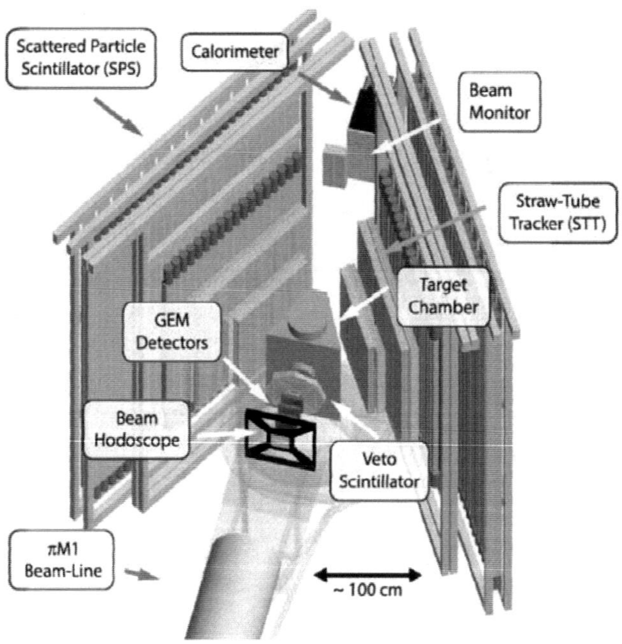

Figure 1. Schematic view of the MUSE setup.

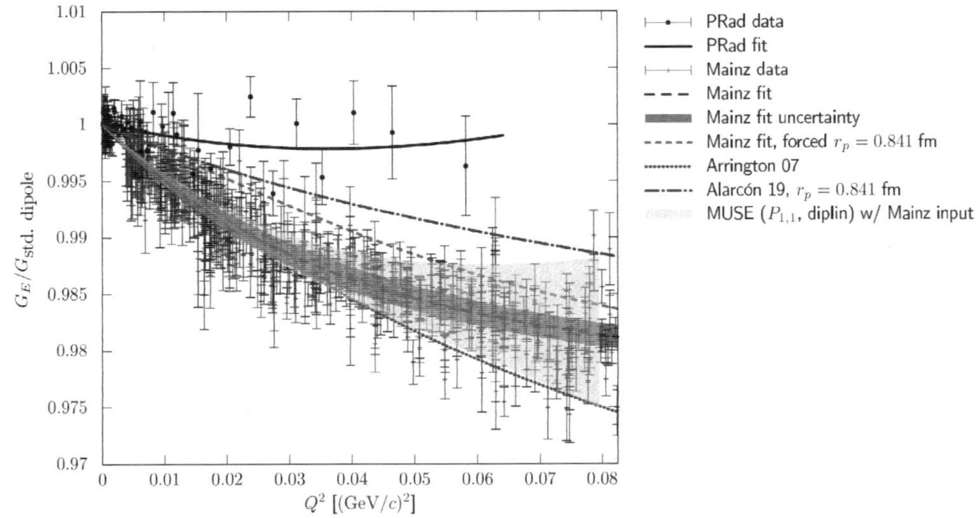

Figure 2. Figure courtesy of J.C. Bernauer. The plot shows the proton electric form factor G_E normalized by the standard dipole $G_{\text{std. dipole}} = (1 + Q^2/0.71)^{-2}$ as a function of Q^2, with data from Mainz [6] and Jefferson Lab [9], and several curves representing a variety of analyses. The projected error band for MUSE is overlaid in yellow.

3. Sensitivity

The MUSE experiment will measure elastic cross sections, extract the electric form factor over a wide range of four-momentum transfer, $Q^2 = 0.002 - 0.07$ (GeV/c)^2, and determine the proton charge radius with four different lepton probes. Figure 2 shows the distribution of the electric form factor G_E, normalized to the dipole form factor, at low Q^2, with data from Mainz [6] and Jefferson Lab [9], and several curves that represent a variety of analyses. Overlaid in yellow is the expected uncertainty band for a future extraction with MUSE data. The precision will be high enough to distinguish between the discrepant data sets from Mainz and Jefferson Lab.

The TPE effects are studied by comparing the yields between the charge polarity of a given lepton family e^+ vs. e^-, and μ^+ vs. μ^-, respectively, shown for muons in Fig. 3, for the planned three values of beam momentum settings [10].

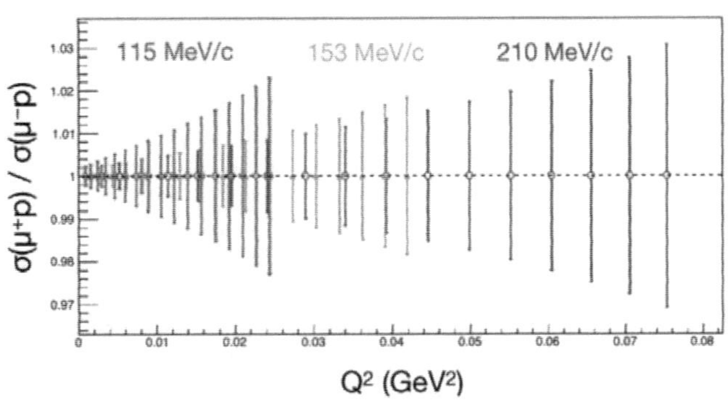

Figure 3. Relative uncertainties in the ratio of μ^+p to μ^-p elastic cross sections, also referred to as muon two-photon exchange (TPE) ratio.

The TPE effect is canceled in the charge average, i.e. precise cross sections, electric form factor distributions, and charge radii are extracted for both lepton species, e and μ. The differences of cross sections, form factors, and charge radii between the e and μ probes are sensitive to lepton non-universality, and are measured with even higher precision than the individual quantitities, since some of the systematic uncertainties are common to both species present in the same beam under the same experimental and kinematic conditions, and partially cancel. The projected uncertainties for lepton universality tests for the three beam momenta are shown in Fig. 4 [10].

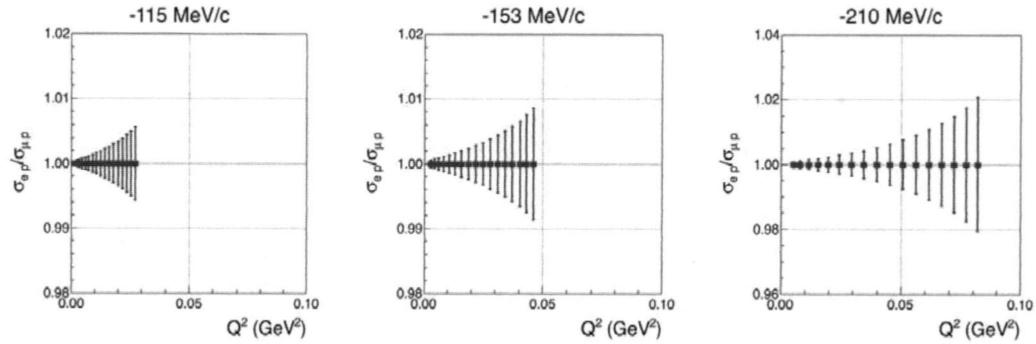

Figure 4. Relative uncertainties in the ratio of μp to ep elastic cross sections, also referred to as lepton universality (LU) ratio.

4. Status and Timeline

The MUSE experiment was first conceived at a workshop in Morocco in discussions between Ron Gilman, Gerald Miller, and M.K. [11] - it was first proposed to the Program Advisory Committee at PSI in 2012 [12] and conditionally approved in 2013 [13]; from 2013-2016 beamline and detectors tests were performed at the piM1 beamline to identify the best design of the MUSE experiment. Funding to construct the apparatus began in 2016 following the approved baseline design documented in the Technical Design Report [10]. Initial construction was completed in 2019 [14, 15]. Commissioning of the full apparatus began in 2019 and has been reiterated through 2022, including a full characterization of the piM1 beamline [8]. In 2021 a first scattering data set was collected. In 2022 the apparatus was further improved by adding an enlarged electromagnetic calorimeter in the downstream region to control radiative corrections, and by adding active veto counters inside the scattering chamber to suppress reading out events originating from chamber-post scattering. Further scattering data taking is beginning in Fall 2022, for a total of 12 beam months over the next two years to acquire the required statistics.

Acknowledgments

This material is based upon work supported by the National Science Foundation under NSF grants PHY-1812402 and PHY-2113436. The MUSE experiment is supported by the Department of Energy, NSF, PSI, and the US-Israel Binational Science Foundation.

5. References

[1] R. Pohl *et al.*, Nature **466**, 213 (2010).
[2] J.C. Bernauer *et al.*, Phys. Rev. Lett. **105**, 242001 (2010).
[3] P.J. Mohr, B.N. Taylor, and D.B. Newell, Rev. Mod. Phys. **84**, 1527 (2012).

[4] P.J. Mohr, D.B. Newell, and B.N. Taylor, Rev. Mod. Phys. **88**, 035009 (2016).
[5] A. Antognini *et al.*, Science **339**, 417 (2013).
[6] J.C. Bernauer *et al.*, Phys. Rev. C. **90**, 015206 (2014).
[7] H. Gao and M. Vanderhaeghen, Rev. Mod. Phys. **94**, 015002 (2022).
[8] E. Cline *et al.*, PRC **105**, 055201 (2022).
[9] W. Xiong *et al.*, Nature **575**, 147 (2019).
[10] R. Gilman *et al.*, *Technical Design Report for the Paul Scherrer Institute Experiment R-12-01.1: Studying the Proton "Radius" Puzzle with μp Elastic Scattering*, `arxiv:1709.09753`.
[11] Int. Workshop on Partons in Nucleons and Nuclei (PINAN11), Marrakech, Morocco, September 26 - 30, 2011.
[12] R. Gilman *et al.*, *Studying the Proton "Radius" Puzzle with μp Elastic Scattering*, Proposal R-12-01.1, BVR43, PSI (2012).
[13] R. Gilman *et al.*, *Studying the Proton "Radius" Puzzle with μp Elastic Scattering*, White Paper on the Paul Scherrer Institut MUSE experiment, `arXiv:1303.2160`.
[14] P. Roy *et al.*, Nucl. Instr. Meth. Phys. Res. A **949**, 162874 (2020).
[15] T. Rostomyan *et al.*, Nucl. Instr. Meth. Phys. Res. A **986**, 164801 (2021).

New Opportunities for the Study of Baryon Number Violation at Low-Energy Accelerators

Susan Gardner

Department of Physics and Astronomy, University of Kentucky, Lexington, KY 40506-0055 USA

E-mail: susan.gardner@uky.edu

Abstract. I motivate new searches for baryon-number violation and consider, particularly, the prospects for detecting baryon number violation by two units at low-energy accelerators with intense electron beams, using ARIEL as a particular example.

1. Why Study Baryon-Number Violation?

The Standard Model (SM) of particle interactions offers no explanation for dark matter or dark energy, nor of nonzero neutrino masses and mixings; and with the measured Higgs mass, it is incompatible with a nonzero cosmic baryon asymmetry (BAU) [1, 2]. Thus each of these observed effects documents the existence of physics beyond the SM. The possibility of detectable baryon-number violation may be a key thread linking the solution to all of these puzzles. Certainly, baryon-number violation has long been considered an essential ingredient to the realization of a nonzero baryon asymmetry: B (baryon number), C (particle-antiparticle), and CP (matter-antimatter) violation must all exist in a non-equilibrium environment for a nonzero BAU to appear [3]. Yet in the SM, the electroweak phase transition is not of first order [1, 2], and baryon-number violation is only expected to occur at exceptionally high temperatures and thus be unobservable today. It is possible, however, that the cosmic origin of dark matter and ordinary baryonic matter are linked [4, 5, 6], with dark matter carrying a conserved charge analogous to baryon number. Perhaps baryon number is conserved across visible and dark sectors [7], suggesting the possibility of nucleon decays to dark final states [8]. Moreover, the existence of fundamental Majorana dynamics is signalled by the breaking of $B - L$ symmetry [9]. Thus through the breaking of L (lepton number) by two units the neutrino can gain an effective Majorana mass and thus generate neutrinoless double β decay, with the existence of the latter speaking to the former [10]. This physics can be studied in the baryon sector as well, with B broken by two units to give rise to $n\bar{n}$ oscillations.

2. Mechanisms of Baryon-Number Violation

Although the experimental limits on proton decay, or, more generally, $|\Delta B| = 1$ processes, are remarkably stringent [11], with partial lifetimes to particular final states in excess of 10^{32} years at 90% CL, processes in which baryon number changes by two (or more) units can be of distinct physical origin [12, 13, 14, 15, 16] and are also much less constrained [15, 16]. *Apparent* baryon-number violation is also possible: the neutron can decay to an invisible final-state that conserves baryon number. Since the final-state particles are undetected, the process makes it appear as if baryon number is broken. This last possibility has gained in interest because of the emergence, over the last decades, of an anomaly in measurements of the neutron lifetime: the lifetime determined from counting the protons that appear, e.g., the beam result $887.7 \pm 1.2\,(\text{stat}) \pm 1.9\,(\text{sys})\,\text{s}$ [17], is significantly different from that determined counting the neutrons that persist, e.g., the bottle result $877.75 \pm 0.28\,(\text{stat})^{+0.22}_{-0.16}\,(\text{sys})\,\text{s}$ [18]. Different explanations have been advanced for this difference; here we consider explanations that invoke hidden sectors with baryon number. The possibility that neutron-"mirror neutron" mixing could exist and be enhanced by the large magnetic fields used in the beam experiment [19], reducing the number of neutrons that decay, has been excluded as an explanation through a direct search [20]. The anomaly could be explained through dark decays of the neutron [21], via $n \to \chi\gamma$ or $n \to \chi e^+e^-$, where the mass of χ must be chosen to evade $|\Delta B| = 1$ and nuclear stability constraints – see Fig. 1 for a concrete example. Constraints on this scenario exist also. Precision tests of the SM charged-weak current [22, 23, 24], as shown in Fig. 2, direct searches [25, 26, 27], and the existence of neutron stars of $2M_\odot$ in mass [28, 29, 30] all constrain the dark decay scenario, though modes such as $n \to 3\chi'$ [31] evade the last two constraints. Generally, the neutron dark decay rate can grossly exceed that of visible $|\Delta B| = 1$ decays by many orders of magnitude [24].

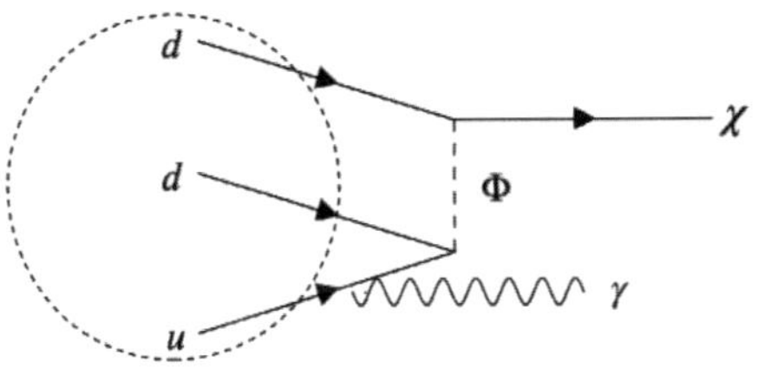

Figure 1. A microscopic, minimal model for $n \to \chi\gamma$ decay requires only a Dirac fermion χ, a SM singlet with $B = 1$, and a color-triplet, weak SU(2) singlet scalar Φ with hypercharge -1/3 and baryon number $-2/3$, as per [21].

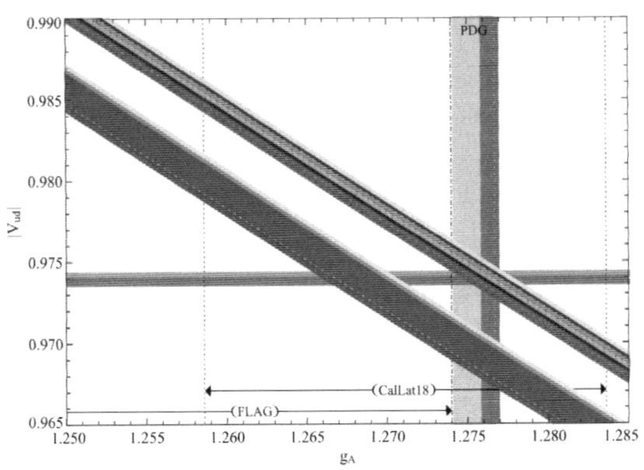

Figure 2. Constraints on the CKM element $|V_{ud}|$ and the axial coupling constant g_A in the SM, with different inputs and estimated radiative corrections (RCs), taken at 68% C.L. – see [24] for all details. Note the beam lifetime [17] (cyan), as well as the 2020 PDG average of bottle/trap lifetimes [32] (purple) and the latest magnetic trap result [33] (green) for different RCs. Values of g_A from lattice QCD are the 2021 FLAG average for $N_f = 2 + 1 + 1$ flavors [34] and the 2018 CalLat result [35]. The decay correlation determinations are from the PDG [32] and from [36] (ochre), with $|V_{ud}|$ from $0^+ \to 0^+$ nuclear β decays (red).

If we suppose that baryon number is broken at energies far above the weak scale, we can make a model-independent analysis of its low-energy breaking pattern through consideration of the new effective operators that appear [37]. Working in the context of the gauge symmetries and particle content of the SM [38, 39], the leading operators with mass dimension $d > 4$ are of the form

$$\mathcal{L}^{(d=6)}_{|\Delta B|=1} \supset \sum_i \frac{c_i}{\Lambda^2_{|\Delta B|=1}} (qqq\ell)_i + \text{h.c.} \tag{1}$$

and

$$\mathcal{L}^{(d=9)}_{|\Delta B|=2} \supset \sum_i \frac{c_i}{\Lambda^5_{|\Delta B|=2}} (qqqqqq)_i + \text{h.c.}\,, \tag{2}$$

for $|\Delta B| = 1$ and $|\Delta B| = 2$ transitions, respectively. Working in the Warsaw basis [39], with $c^{(\ell)}_{\chi\chi'} = 1$ and allowing one operator to act at a time, ultimately yields the constraints on $|\Delta B| = 1$ operators shown in Fig. 3. Thus we see that the scale of new physics in this case satisfies $\Lambda_{|\Delta B|=1} > 10^{14}$ GeV. In contrast, $|\Delta B| = 2$ transitions are engendered by higher mass dimension

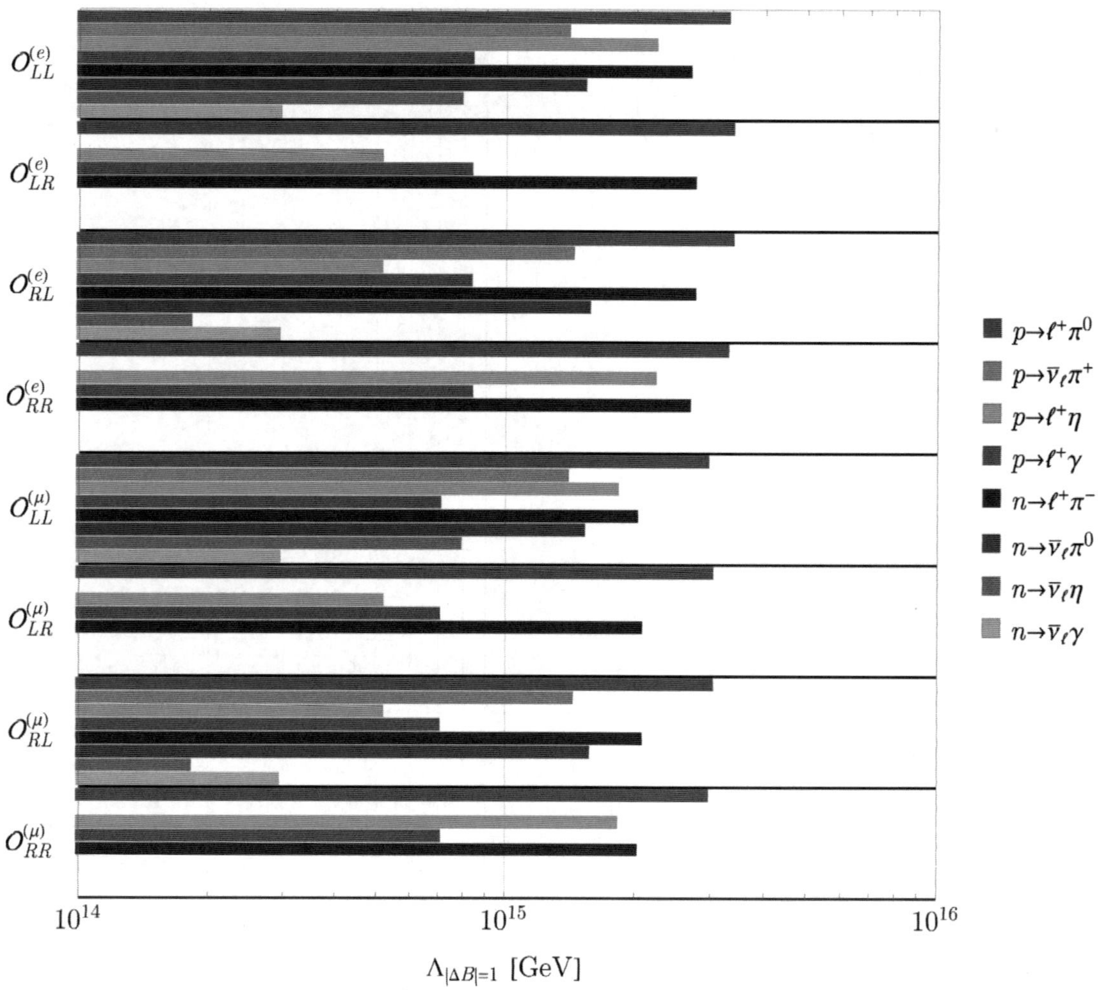

Figure 3. Constraints on individual $|\Delta B| = 1$ operators from experimental limits at 90% C.L., from [24]. The operators $\mathcal{O}^{(\ell)}_{\chi R}$ for $\chi \in L, R$ are not constrained by decays with antineutrino final states. Evading the different empirical constraints through the cancellation of different operator contributions is not tenable.

operators, so that we would expect a lower new-physics scale in this case. Using the experimental limit on $n\bar{n}$ oscillations [40, 41], a simple estimate yields $\Lambda_{|\Delta B|=2} \gtrsim 10^{5.5}$ GeV. Establishing the existence of $n - \bar{n}$ oscillations is of great importance, because it would demonstrate that fundamental Majorana dynamics exists in Nature, that the neutron also has a Majorana mass. Such a mass term is possible for any electrically neutral massive fermion, but the neutron's nonzero magnetic moment means that a Majorana neutron would be an entangled n and $\bar{n}$ state, whereas a Majorana neutrino can be a two-component field [42]. Nucleon-antinucleon transitions can be realized in different ways: through (i) searches for neutron-antineutron oscillations of free neutrons and in nuclei, (ii) searches for dinucleon decays in nuclei, and (iii) searches for nucleon-antinucleon conversion as mediated through a low-energy scattering process [43, 15]. The oscillation of a neutron into an antineutron would be a spontaneous process, and its likelihood is expected to be strongly suppressed by environmental effects that distinguish neutrons and antineutrons [44] – thus new experiments would be conducted under high vacuum with extreme magnetic field mitigation [45, 46]. Searches for dinucleon decays are limited by the finite density of the nucleus; if the BSM mediators are of much shorter range than the NN force, then dinucleon decays can be strongly suppressed by the likelihood of finding two nucleons sufficiently close together [15]. Here we focus on the last possibility, particularly that of $|\Delta B| = 2$ processes mediated by low-energy electron scattering [43, 15]. We find particular motivation for such studies in the connection they can provide between $n - \bar{n}$ oscillations and the short-range mechanism of neutrinoless double β decay, which we illustrate in Fig. 4. For

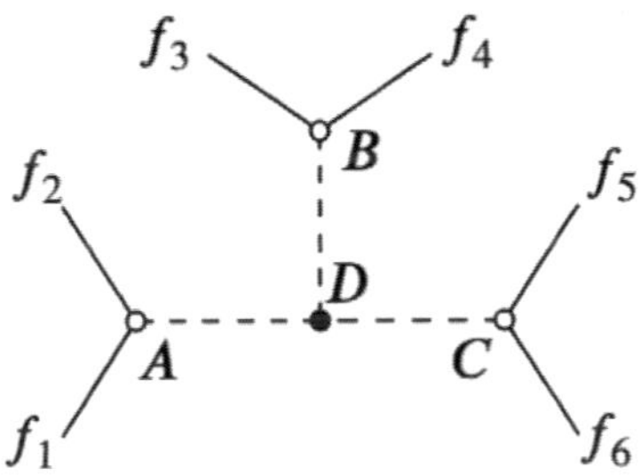

Figure 4. A short-range mechanism exploiting new $B-L$ violating dynamics to realize either $n-\bar{n}$ oscillations or neutrinoless double β decay, with the choices differentiated by the explicit fermion content $f_{1,...6}$ and new scalar or vector mediators that carry either B or L as appropriate to the context. It is possible to assert the existence of one given the observation of the other, if a $B - L$ conserving $|\Delta B| = 2$ process is also observed [15], and connecting the two via the observation of a $|\Delta B| = 1$ process is also possible [47, 24].

context, we contrast different mechanisms of neutrinoless double β decay in Fig. 5; their relative size depends on the mechanism of new physics [48].

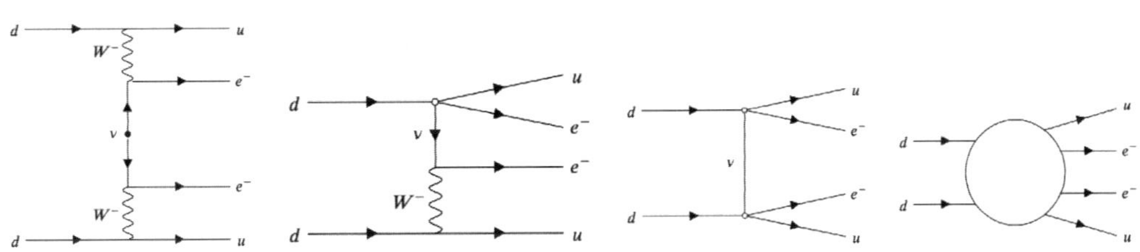

Figure 5. Mechanisms of neutrinoless double β decay, at the quark level, after [49]. The solid dot indicates a Majorana neutrino mass and hence the "mass mechanism," with open circles indicating effective interactions generated by new physics at a higher energy scale. The rightmost graph is the "short range" mechanism also shown in Fig. 4; within nuclei this can appear as $\pi^-\pi^- \to e^-e^-$ decay [48]. Analyses within effective field theory also exist [50, 51].

3. Modelling Baryon-Number Violation without Proton Decay

Given the severity of the constraints associated with $|\Delta B| = 1$ processes [24], we choose to model processes with $|\Delta B| = 2$ and/or $|\Delta L = 2|$ within minimal scalar models without proton decay [13, 14, 15]. That is, we consider all the models with renormalizable and SM gauge-invariant scalar-fermion interactions that do not mediate tree-level $|\Delta B| = 1$ processes, generalizing [13]. We enumerate the possible new scalars X_i in Table 1. The $(3, 1, -1/3)$ scalar in the model of Fig. 1 [21] does not appear because it mediates proton decay. With this set of X_i, B- and L- violating interactions appear through their mutual interactions, constructed to be of SM gauge invariant form: $X_iX_jX_k$ or $X_iX_jX_kX_l$. Since the interactions are all of mass dimension 4 or less, the associated masses and couplings are to be limited by experiment [15], very much in the spirit of "hidden sector" searches for dark photons and particles of similar ilk at intermediate-energy electron accelerators and elsewhere, note, e.g., Fig. 3.3 of [52]. The possible realizations of $|\Delta B| = 2$ and/or $|\Delta L| = 2$ violations that result are given in Table 2. Interactions with 4 X_i do not break $B - L$, but ones with 3 X_i do, as per SM expectations [53]. The interactions for a particular choice of X_i are

$$\mathcal{L}_{\rm int} \supset -g_1^{ab} X_1 (e^a e^b) - g_7^{ab} X_7^{\alpha\beta} (d_\alpha^a d_\beta^b) - g_8^{ab} X_8^{\alpha\beta} (u_\alpha^a u_\beta^b) - \lambda_3 X_7^{\alpha\alpha'} X_7^{\beta\beta'} X_8^{\gamma\gamma'} \epsilon_{\alpha\beta\gamma}\epsilon_{\alpha'\beta'\gamma'} \\ -\lambda_{10} X_7^{\alpha\alpha'} X_8^{\beta\beta'} X_8^{\gamma\gamma'} X_1 \epsilon_{\alpha\beta\gamma}\epsilon_{\alpha'\beta'\gamma'} - \lambda_A X_8^{\alpha\alpha'} (X_7^{\alpha\alpha'})^\dagger X_1 + \text{h.c.} \quad (3)$$

Although each term has mass dimension 4 or less, at low energies $\mathcal{L}_{\rm int}$ generates a $d = 9$ operator to yield $n\bar{n}$ oscillations, a $d = 12$ operator to yield $e^-p \to e^+\bar{p}$, and a $d = 9$ operator to yield $\pi^-\pi^- \to e^-e^-$ decay, i.e., a short-distance contribution to $0\nu\beta\beta$ decay. Feynman diagrams for these processes, working at energies for which X_1, X_7, X_8 can appear explicitly, are shown in Fig. 6. There are two different possible models, containing different X_i, that assert that $\pi^-\pi^- \to e^-e^-$ should exist if both $n\bar{n}$ oscillations and $e^-p \to e^+\bar{p}$ are observed, and they are distinguished by whether $e^-p \to \bar{\nu}_X\bar{n}$, where the neutrino flavor is undetermined, or $e^-n \to e^-\bar{n}$ are also observed [15]. Thus high-intensity, low-energy electron scattering facilities, which may be under development to test new accelerator concepts [54], can also be used to broader purpose [43, 15]. In what follows we focus on the detectability of $e^-p \to e^+\bar{p}$ and $e^-p \to \bar{\nu}_X\bar{n}$ in that context [55]. We note [16, 24] for broader discussions of the interconnections between processes that break $B - L$ by two or units.

Table 1. B- or L-carrying scalars that permit no proton decay at tree level, after [13, 15]. Possible scalar-fermion interactions are indicated schematically, with L-handed doublets L^a, Q^a and R-handed singlets e^a, u^a, d^a and generational indices a, b, noting the symmetry of the associated scalar-fermion couplings under $a \leftrightarrow b$ exchange. Note $Q_{\rm em} = T_3 + Y$.

Scalar	$SU(3)_c \times SU(2)_L \times U(1)_Y$	B	L	Operator(s)	$[g_i^{ab}?]$
X_1	$(1, 1, 2)$	0	-2	Xe^ae^b	[S]
X_2	$(1, 1, 1)$	0	-2	XL^aL^b	[A]
X_3	$(1, 3, 1)$	0	-2	XL^aL^b	[S]
X_4	$(\bar{6}, 3, -1/3)$	-2/3	0	XQ^aQ^b	[S]
X_5	$(\bar{6}, 1, -1/3)$	-2/3	0	XQ^aQ^b, Xu^ad^b	[A,–]
X_6	$(3, 1, 2/3)$	-2/3	0	Xd^ad^b	[A]
X_7	$(\bar{6}, 1, 2/3)$	-2/3	0	Xd^ad^b	[S]
X_8	$(\bar{6}, 1, -4/3)$	-2/3	0	Xu^au^b	[S]
X_9	$(3, 2, 7/6)$	1/3	-1	$X\bar{Q}^ae^b, XL^a\bar{u}^b$	[–,–]

Journal of Physics: Conference Series **2391** (2022) 012016 doi:10.1088/1742-6596/2391/1/012016

Table 2. Minimal interactions of the scalars X_i of Table 1, indicated schematically, after [15]. Interactions labelled M1-M9 appear as models 1-9 in [13]. Interactions M1-M4 have $|\Delta B| = 2$, $|\Delta L| = 0$, whereas A-G have $|\Delta L| = 2$, $|\Delta B| = 0$. Further models follow from M8, M17, and M18 under $X_7 \to X_6$, but they cannot involve first-generation fermions only.

Model		Model		Model	
M1	$X_5X_5X_7$	A	$X_1X_8X_7^\dagger$	M10	$X_7X_8X_8X_1$
M2	$X_4X_4X_7$	B	$X_3X_4X_7^\dagger$	M11	$X_5X_5X_4X_3$
M3	$X_7X_7X_8$	C	$X_3X_8X_4^\dagger$	M12	$X_5X_5X_8X_1$
M4	$X_6X_6X_8$	D	$X_5X_2X_7^\dagger$	M13	$X_4X_4X_5X_2$
M5	$X_5X_5X_5X_2$	E	$X_8X_2X_5^\dagger$	M14	$X_4X_4X_5X_3$
M6	$X_4X_4X_4X_2$	F	$X_2X_2X_1^\dagger$	M15	$X_4X_4X_8X_1$
M7	$X_4X_4X_4X_3$	G	$X_3X_3X_1^\dagger$	M16	$X_4X_7X_8X_3$
M8	$X_7X_7X_7X_1^\dagger$			M17	$X_5X_7X_7X_2^\dagger$
M9	$X_6X_6X_6X_1^\dagger$			M18	$X_4X_7X_7X_3^\dagger$

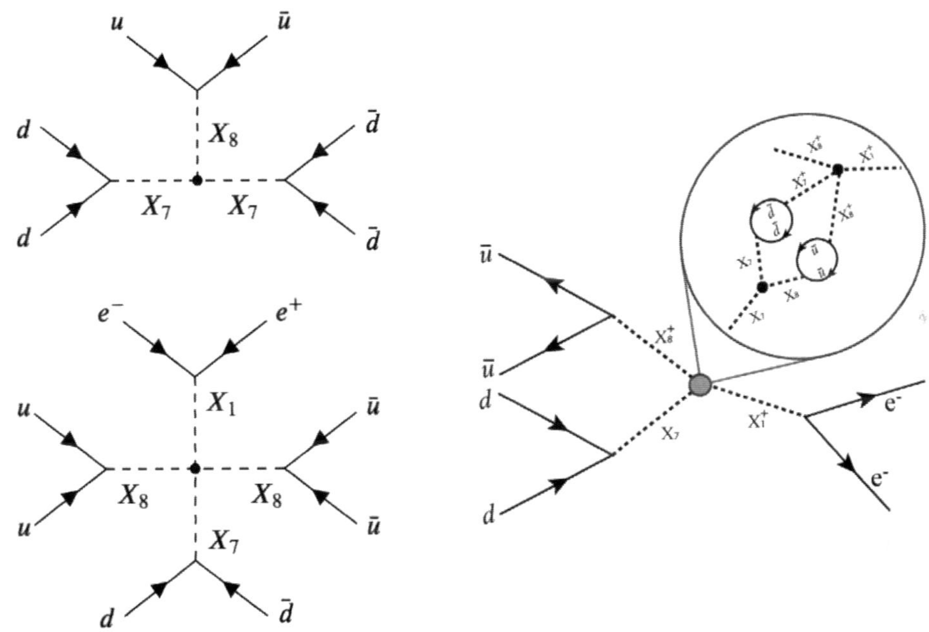

Figure 6. Connecting $n\bar{n}$ and $e^-p \to e^+\bar{p}$ to $\pi^-\pi^- \to e^-e^-$: if the first two processes are established, then the third must exist also [15], because anything not forbidden is compulsory [56].

4. Phenomenology of New Scalars: Existing Constraints

The new scalars we have noted find constraints from many sources [15, 55]. If the scalars couple to all three generations, then constraints from flavor physics, particularly from the $|\Delta F| = 2$ FCNCs and the neutron EDM limit, limit the different generational couplings significantly even if the scalar masses were well in excess of a TeV [13, 57, 58]. Here we suppose, rather, that the scalars couple to first-generation fermions only, so that these constraints do not apply. Although

Journal of Physics: Conference Series **2391** (2022) 012016 doi:10.1088/1742-6596/2391/1/012016

one might think that the new scalars, particularly the non-color-singlet ones, must nonetheless be of $\mathcal{O}(1\,\text{TeV})$ in mass [59], regions of parameter space associated with sub-$\mathcal{O}(10\,\text{GeV})$ masses are not yet excluded [15]. It is difficult to challenge this mass region at colliders due to QCD backgrounds [60], yet the new scalars all carry electric charge – and additional ingredients are required to negate would-be contributions to the width of the Z^0 gauge boson [61]. With these in place, we have constraints from the following sources, namely, from (i) the non-observation of $n-\bar{n}$ oscillations, though some model choices would not produce it, (ii) direct searches for same-sign dileptons at colliders, though to control backgrounds same-sign lepton pairs of the same flavor with an invariant mass of less than 8 GeV are rejected [62, 63], (iii) nuclear stability studies, particularly the limit on $pp \to e^+e^+$ and related modes from ^{16}O [64, 65], (iv) non-observation of GeV-scale gamma rays from hydrogen-antihydrogen ($\mathrm{H}\bar{\mathrm{H}}$) annihilation in hydrogen clouds [66, 67], and, finally, (v) indirect constraints from measurements of $(g-2)_e$ [68, 69, 61] and from parity-violating Møller scattering [70, 71, 61]. Both (iii) and (iv) stem from the non-observation of spontaneous processes which would conceivably occur if the needed new physics were present. However, conventional physics can act to make a spontaneous process impossible, in that it can (1) forbid a process through energy-momentum conservation or (ii) make the required initial configuration of particles impossible to realize, or it can negate the appearance of the supposed final state. The first possibility is long known in the hunt for free neutron-antineutron oscillations, requiring that searches be conducted in regions free from magnetic fields or from other matter [45, 46]. We note that Coulomb repulsion and binding-energy effects grossly reduce new-physics limits from ^{16}O stability studies, and galactic magnetic fields [72] weaken the ability of $\mathrm{H}\bar{\mathrm{H}}$ oscillations to occur [15]. The *Fermi* LAT collaboration has also noted a dearth of *expected* GeV-scale gamma rays, which they interpret as evidence for intergalactic magnetic fields [73]. The indirect limits of (v) constrain the scalar-electron couplings, with the most severe limit coming from the Moller E158 experiment [70], yielding $M_{X_{1,3}}/g_{1,3}^{11} \geq 2.7\,\text{TeV}$ at 90% CL for $M_{X_{1,3}} > 0.23\,\text{GeV}$ [61]. Employing this constraint we consider $0.23\,\text{GeV} < M_{X_{1,3}} < 8\,\text{GeV}$ and the colored scalars of Table 1 with masses between roughly $1-10\,\text{GeV}$ as viable possibilities.

5. Probing Baryon-Number Violation by 2 units at ARIEL

To compute $e^-p \to e^+\bar{p}$ or $e^-p \to \bar{\nu}\bar{n}$, e.g., we need to evaluate the Feynman diagram of $4X$ form shown in Fig. 6. In what follows I summarize the work of an upcoming paper with Xinshuai Yan [55]. At ARIEL energies, the new scalars cease to be dynamical degrees of freedom, and we can match that description to an effective theory in which a new 6-fermion contact interaction appears, noting that its prefactor has terms that scale as $(g_i^{11})^2/M_{X_i}^2$. Finally we can match that description to an effective hadron theory, in which the matching coefficient of the $e_\chi^T C e_\chi N_{\chi'}^T C p_{\chi'}$ effective operator is determined by the hadronic matrix element of the 4-quark operator computed in the M.I.T. bag model [74]. Since the momentum transfer in the process is small, i.e., $Q^2 \ll M_{X_i}^2$, we set $Q^2 = 0$ in our numerical assessment. Such bag-model estimates have been shown to compare fairly well with lattice QCD computations for $n\bar{n}$ oscillations [43]. Thus, finally, we can compute the cross sections – and event rates – we would expect for the various nucleon-antinucleon conversion processes of interest, mediated via the different scalar interaction models, as a function of the new scalar masses and couplings. Thus we can work beyond the naive dimensional analysis estimates previously employed [13, 15, 59].

Turning to the rate estimates, we choose $M_{X_i}/(|g_i^{11}|)^{1/2} = 30\,\text{GeV}$ for $i \in 1-3$ and $M_{X_i}/(|g_i^{11}|)^{1/2} = 1\,\text{GeV}$ for $i \in 4-8$ with $\lambda_j = 1$ for all j. We employ the e-linac beam mode at ARIEL [75] with a current of 10 mA and a beam energy of 40 MeV and a 100 cm long hydrogen target with density of $0.09 \times 10^{-3}\,\text{g/cm}^{-3}$ and assume an antinucleon detection efficiency of 100%. Thus we estimate that model M10, yielding $e^-p \to e^+\bar{p}$, would give about 13 events per year of running [55], with the various possible models giving outcomes roughly ranging from less than 1 event per year to about 150 events per year [55].

We emphasize that such rare processes can be identified in the environment of a low-energy, high-intensity electron scattering facility. The electron beam energy at ARIEL is far below pion-production threshold, so that the expected annihilation signature of a multi-pion final state should be striking. In a $n-\bar{n}$ experiment with cold neutrons such a signal is thought to be background free [45, 46]. We refer to [76] for an annihilation model for both free $\bar{p}p$ annihilation at rest and for $\bar{p}$ annihilation in complex targets, noting that the free proton model agrees favorably with experiment [77, 78] and the most likely outcome is a five-pion final state [79, 76]. Low-energy $\bar{n}p$ annihilation data exist as well, and although they appear to suggest a strong isoscalar component [80], a $\bar{p}p$-based analysis appears to give a reasonable indication of the expected annihilation outcomes [76]. For the low-energy processes we consider at ARIEL, we expect the annihilation process to be prompt and likely within the volume of the target.

6. Summary

We have considered the motivations for the study of baryon-number violation at low energies and have shown that new, possible avenues for B and/or L violation by 2 units, or more, have been largely overlooked. These studies can provide new insights into the nature of the neutrino. New scalars that could help mediate rare B and/or L-violating processes are associated with dimension $d \geq 9$ effective operators and are not excluded by existing experiments. Consequently we have considered the discovery potential of possible new experiments such as $e^- p \to e^+ \bar{p}$ or $e^- p \to \bar{\nu} \bar{n}$. In a hydrogen target, the two possibilities should be distinguishable by the net electric charge of the produced multi-pion state. These possibilities could be fruitfully explored at an intense, low-energy electron accelerator such as ARIEL, and such prospects strengthen interest in $|\Delta B| = 2$ experiments of increased sensitivity.

Acknowledgments

S.G. thanks Xinshuai Yan for his collaboration on accelerator-based studies of baryon-number violation, highlighted here, and Jeffrey Berryman and Mohammedreza Zakeri for their collaboration on broader studies of baryon-number violation, particularly in neutron stars. I also acknowledge partial support from the U.S. Department of Energy under contract DE-FG02-96ER40989.

References

[1] K. Kajantie, M. Laine, K. Rummukainen, and Mikhail E. Shaposhnikov. Is there a hot electroweak phase transition at $m_H \gtrsim m_W$? *Phys. Rev. Lett.*, 77:2887–2890, 1996.
[2] Y. Aoki, F. Csikor, Z. Fodor, and A. Ukawa. The Endpoint of the first order phase transition of the SU(2) gauge Higgs model on a four-dimensional isotropic lattice. *Phys. Rev. D*, 60:013001, 1999.
[3] A. D. Sakharov. Violation of CP Invariance, C asymmetry, and baryon asymmetry of the universe. *Pisma Zh. Eksp. Teor. Fiz.*, 5:32–35, 1967.
[4] P. Hut and Keith A. Olive. A Cosmological Upper Limit on the Mass of Heavy Neutrinos. *Phys. Lett. B*, 87:144–146, 1979.
[5] S. Nussinov. Technocosmology: could a technibaryon excess provide a 'natural' missing mass candidate? . *Phys. Lett. B*, 165:55–58, 1985.
[6] Stephen M. Barr, R. Sekhar Chivukula, and Edward Farhi. Electroweak Fermion Number Violation and the Production of Stable Particles in the Early Universe. *Phys. Lett. B*, 241:387–391, 1990.
[7] Hooman Davoudiasl and Rabindra N. Mohapatra. On Relating the Genesis of Cosmic Baryons and Dark Matter. *New J. Phys.*, 14:095011, 2012.
[8] Hooman Davoudiasl. Nucleon Decay into a Dark Sector. *Phys. Rev. Lett.*, 114(5):051802, 2015.
[9] Steven Weinberg. Baryon and Lepton Nonconserving Processes. *Phys. Rev. Lett.*, 43:1566–1570, 1979.
[10] J. Schechter and J. W. F. Valle. Neutrino Masses in SU(2) x U(1) Theories. *Phys. Rev. D*, 22:2227, 1980.
[11] R. L. Workman and Others. Review of Particle Physics. *PTEP*, 2022:083C01, 2022.

[12] Rabindra N. Mohapatra and R. E. Marshak. Local B-L Symmetry of Electroweak Interactions, Majorana Neutrinos and Neutron Oscillations. *Phys. Rev. Lett.*, 44:1316–1319, 1980. [Erratum: Phys.Rev.Lett. 44, 1643 (1980)].
[13] Jonathan M. Arnold, Bartosz Fornal, and Mark B. Wise. Simplified models with baryon number violation but no proton decay. *Phys. Rev. D*, 87:075004, 2013.
[14] P. S. Bhupal Dev and Rabindra N. Mohapatra. TeV scale model for baryon and lepton number violation and resonant baryogenesis. *Phys. Rev. D*, 92(1):016007, 2015.
[15] Susan Gardner and Xinshuai Yan. Processes that break baryon number by two units and the Majorana nature of the neutrino. *Phys. Lett. B*, 790:421–426, 2019.
[16] Julian Heeck and Volodymyr Takhistov. Inclusive Nucleon Decay Searches as a Frontier of Baryon Number Violation. *Phys. Rev. D*, 101(1):015005, 2020.
[17] A. T. Yue, M. S. Dewey, D. M. Gilliam, G. L. Greene, A. B. Laptev, J. S. Nico, W. M. Snow, and F. E. Wietfeldt. Improved Determination of the Neutron Lifetime. *Phys. Rev. Lett.*, 111(22):222501, 2013.
[18] F. M. Gonzalez et al. Improved Neutron Lifetime Measurement with UCNτ. *Phys. Rev. Lett.*, 127(16):162501, 2021.
[19] Zurab Berezhiani. Neutron lifetime puzzle and neutron–mirror neutron oscillation. *Eur. Phys. J. C*, 79(6):484, 2019.
[20] L. J. Broussard et al. Experimental Search for Neutron to Mirror Neutron Oscillations as an Explanation of the Neutron Lifetime Anomaly. *Phys. Rev. Lett.*, 128(21):212503, 2022.
[21] Bartosz Fornal and Benjamín Grinstein. Dark Matter Interpretation of the Neutron Decay Anomaly. *Phys. Rev. Lett.*, 120(19):191801, May 2018.
[22] Andrzej Czarnecki, William J. Marciano, and Alberto Sirlin. Neutron Lifetime and Axial Coupling Connection. *Phys. Rev. Lett.*, 120(20):202002, 2018.
[23] D. Dubbers, H. Saul, B. Märkisch, T. Soldner, and H. Abele. Exotic decay channels are not the cause of the neutron lifetime anomaly. *Phys. Lett. B*, 791:6–10, 2019.
[24] Jeffrey M. Berryman, Susan Gardner, and Mohammadreza Zakeri. Neutron Stars with Baryon Number Violation, Probing Dark Sectors. *Symmetry*, 14(3):518, 2022.
[25] Z. Tang et al. Search for the Neutron Decay n$\rightarrow$ X+γ where X is a dark matter particle. *Phys. Rev. Lett.*, 121(2):022505, 2018.
[26] X. Sun et al. Search for neutron dark decay: n $\rightarrow \chi + e^+e^-$. *EPJ Web Conf.*, 219:05008, 2019.
[27] M. Klopf, E. Jericha, B. Märkisch, H. Saul, T. Soldner, and H. Abele. Constraints on the Dark Matter Interpretation $n \rightarrow \chi + e^+e^-$ of the Neutron Decay Anomaly with the PERKEO II experiment. *Phys. Rev. Lett.*, 122(22):222503, 2019.
[28] David McKeen, Ann E. Nelson, Sanjay Reddy, and Dake Zhou. Neutron Stars Exclude Light Dark Baryons. *Phys. Rev. Lett.*, 121(6):061802, August 2018.
[29] Gordon Baym, D. H. Beck, Peter Geltenbort, and Jessie Shelton. Testing Dark Decays of Baryons in Neutron Stars. *Phys. Rev. Lett.*, 121(6):061801, August 2018.
[30] T. F. Motta, P. A. M. Guichon, and A. W. Thomas. Implications of Neutron Star Properties for the Existence of Light Dark Matter. *J. Phys. G*, 45(5):05LT01, 2018.
[31] Alessandro Strumia. Dark Matter interpretation of the neutron decay anomaly. *JHEP*, 02:067, 2022.
[32] P. A. Zyla et al. Review of Particle Physics. *PTEP*, 2020(8):083C01, 2020.
[33] D. González, D. León, B. Fabela, and M. I. Pedraza. Detecting physics beyond the Standard Model with the REDTOP experiment. *J. Phys. Conf. Ser.*, 912(1):012042, 2017.
[34] Y. Aoki et al. FLAG Review 2021, 11 2021.
[35] C. C. Chang et al. A percent-level determination of the nucleon axial coupling from quantum chromodynamics. *Nature*, 558(7708):91–94, 2018.
[36] B. Märkisch et al. Measurement of the Weak Axial-Vector Coupling Constant in the Decay of Free Neutrons Using a Pulsed Cold Neutron Beam. *Phys. Rev. Lett.*, 122(24):242501, 2019.
[37] Thomas Appelquist and J. Carazzone. Infrared Singularities and Massive Fields. *Phys. Rev. D*, 11:2856, 1975.
[38] W. Buchmuller and D. Wyler. Effective Lagrangian Analysis of New Interactions and Flavor Conservation. *Nucl. Phys. B*, 268:621–653, 1986.
[39] B. Grzadkowski, M. Iskrzynski, M. Misiak, and J. Rosiek. Dimension-Six Terms in the Standard Model Lagrangian. *JHEP*, 10:085, 2010.
[40] M. Baldo-Ceolin et al. A New experimental limit on neutron - anti-neutron oscillations. *Z. Phys. C*, 63:409–416, 1994.
[41] K. Abe et al. The Search for $n - \bar{n}$ oscillation in Super-Kamiokande I. *Phys. Rev. D*, 91:072006, 2015.
[42] Susan Gardner and Xinshuai Yan. CPT, CP, and C transformations of fermions, and their consequences, in theories with B-L violation. *Phys. Rev. D*, 93(9):096008, 2016.

[43] Susan Gardner and Xinshuai Yan. Phenomenology of neutron-antineutron conversion. *Phys. Rev. D*, 97(5):056008, 2018.
[44] R. E. Marshak and Rabindra N. Mohapatra. Quark - Lepton Symmetry and B-L as the U(1) Generator of the Electroweak Symmetry Group. *Phys. Lett. B*, 91:222–224, 1980.
[45] D. G. Phillips, II et al. Neutron-Antineutron Oscillations: Theoretical Status and Experimental Prospects. *Phys. Rept.*, 612:1–45, 2016.
[46] A. Addazi et al. New high-sensitivity searches for neutrons converting into antineutrons and/or sterile neutrons at the HIBEAM/NNBAR experiment at the European Spallation Source. *J. Phys. G*, 48(7):070501, 2021.
[47] K. S. Babu and Rabindra N. Mohapatra. Determining Majorana Nature of Neutrino from Nucleon Decays and $n - \bar{n}$ oscillations. *Phys. Rev. D*, 91(1):013008, 2015.
[48] J. D. Vergados. Pion Double Charge Exchange Contribution to Neutrinoless Double Beta Decay. *Phys. Rev. D*, 25:914, 1982.
[49] Florian Bonnet, Martin Hirsch, Toshihiko Ota, and Walter Winter. Systematic decomposition of the neutrinoless double beta decay operator. *JHEP*, 03:055, 2013. [Erratum: JHEP 04, 090 (2014)].
[50] Gary Prezeau, M. Ramsey-Musolf, and Petr Vogel. Neutrinoless double beta decay and effective field theory. *Phys. Rev. D*, 68:034016, 2003.
[51] V. Cirigliano, W. Dekens, J. de Vries, M. L. Graesser, and E. Mereghetti. A neutrinoless double beta decay master formula from effective field theory. *JHEP*, 12:097, 2018.
[52] Marco Fabbrichesi, Emidio Gabrielli, and Gaia Lanfranchi. *The Physics of the Dark Photon*. Springer, 2021.
[53] Andrew Kobach. Baryon Number, Lepton Number, and Operator Dimension in the Standard Model. *Phys. Lett. B*, 758:455–457, 2016.
[54] Emilio A. Nanni et al. C^3 Demonstration Research and Development Plan. In *2022 Snowmass Summer Study*, 3 2022.
[55] Susan Gardner and Xinshuai Yan. Prospects for the accelerator-based discovery of baryon-number violation by two units. to appear on the arXiv.
[56] M. Gell-Mann. The interpretation of the new particles as displaced charge multiplets. *Nuovo Cim.*, 4(S2):848–866, 1956.
[57] Elaine C. F. S. Fortes, K. S. Babu, and R. N. Mohapatra. Flavor Physics Constraints on TeV Scale Color Sextet Scalars. In *6th International Workshop on Charm Physics*, 11 2013.
[58] K. Babu et al. Neutron-Antineutron Oscillations: A Snowmass 2013 White Paper, 10 2013.
[59] Andreas Helset, Clara Murgui, and Mark B. Wise. Simple models with both baryon and lepton number violation by two units. *Phys. Rev. D*, 104(1):015029, 2021.
[60] Peter Richardson and David Winn. Simulation of Sextet Diquark Production. *Eur. Phys. J. C*, 72:1862, 2012.
[61] Susan Gardner and Xinshuai Yan. Light scalars with lepton number to solve the $(g-2)_e$ anomaly. *Phys. Rev. D*, 102(7):075016, 2020.
[62] Serguei Chatrchyan et al. Search for New Physics in Events with Same-Sign Dileptons and Jets in pp Collisions at $\sqrt{s} = 8$ TeV. *JHEP*, 01:163, 2014. [Erratum: JHEP 01, 014 (2015)].
[63] Vardan Khachatryan et al. Search for new physics in same-sign dilepton events in proton–proton collisions at $\sqrt{s} = 13\,\text{TeV}$. *Eur. Phys. J. C*, 76(8):439, 2016.
[64] S. Sussman et al. Dinucleon and Nucleon Decay to Two-Body Final States with no Hadrons in Super-Kamiokande, 11 2018.
[65] Joseph Bramante, Jason Kumar, and John Learned. Proton annihilation at hadron colliders and Kamioka: high-energy versus high-luminosity. *Phys. Rev. D*, 91(3):035012, 2015.
[66] A. A. Abdo et al. Fermi LAT Observation of Diffuse Gamma-Rays Produced Through Interactions between Local Interstellar Matter and High Energy Cosmic Rays. *Astrophys. J.*, 703:1249–1256, 2009.
[67] Yuval Grossman, Wee Hao Ng, and Shamayita Ray. Revisiting the bounds on hydrogen-antihydrogen oscillations from diffuse γ-ray surveys. *Phys. Rev. D*, 98(3):035020, 2018.
[68] D. Hanneke, S. Fogwell, and G. Gabrielse. New Measurement of the Electron Magnetic Moment and the Fine Structure Constant. *Phys. Rev. Lett.*, 100:120801, 2008.
[69] K. S. Babu and C. Macesanu. Two loop neutrino mass generation and its experimental consequences. *Phys. Rev. D*, 67:073010, 2003.
[70] P. L. Anthony et al. Precision measurement of the weak mixing angle in Moller scattering. *Phys. Rev. Lett.*, 95:081601, 2005.
[71] P. S. Bhupal Dev, Michael J. Ramsey-Musolf, and Yongchao Zhang. Doubly-Charged Scalars in the Type-II Seesaw Mechanism: Fundamental Symmetry Tests and High-Energy Searches. *Phys. Rev. D*, 98(5):055013, 2018.
[72] Richard M. Crutcher, Benjamin Wandelt, Carl Heiles, Edith Falgarone, and Thomas H. Troland. Magnetic

Fields in Interstellar Clouds from Zeeman Observations: Inference of Total Field Strengths by Bayesian Analysis. *Ap. J.*, 725(1):466–479, December 2010.

[73] A. Neronov and I. Vovk. Evidence for strong extragalactic magnetic fields from Fermi observations of TeV blazars. *Science*, 328:73–75, 2010.

[74] A. Chodos, R. L. Jaffe, K. Johnson, Charles B. Thorn, and V. F. Weisskopf. A New Extended Model of Hadrons. *Phys. Rev. D*, 9:3471–3495, 1974.

[75] Jonathan Bagger, Friedhelm Ames, Yu. Bylinskii, Alexander Gottberg, Kathryn Hayashi, Oliver Kester, Shane Koscielniak, Robert Laxdal, Marco Marchetto, and Paul Schaffer. ARIEL at TRIUMF: Science and Technology. In *9th International Particle Accelerator Conference*, 6 2018.

[76] Joshua L. Barrow, Elena S. Golubeva, Eduard Paryev, and Jean-Marc Richard. Progress and simulations for intranuclear neutron-antineutron transformations in $^{40}_{18}Ar$. *Phys. Rev. D*, 101(3):036008, 2020.

[77] P. Salvini et al. $\bar{p}p$ annihilation into four charged pions at rest and in flight. *Eur. Phys. J. C*, 35:21–33, 2004.

[78] C. Amsler et al. Annihilation at rest of antiprotons and protons into neutral particles. *Nucl. Phys. A*, 720:357–367, 2003.

[79] Eberhard Klempt, Chris Batty, and Jean-Marc Richard. The antinucleon-nucleon interaction at low energy : Annihilation dynamics. *Phys. Rept.*, 413:197–317, 2005.

[80] F. Iazzi et al. Antineutron proton total cross-section from 50-MeV/c to 400-MeV/c. *Phys. Lett. B*, 475:378–385, 2000.

Journal of Physics: Conference Series **2391** (2022) 012017 doi:10.1088/1742-6596/2391/1/012017

Proton Radius: A Puzzle or a Solution!?

Ulrich D. Jentschura

Department of Physics and LAMOR, Missouri University of Science and Technology, Rolla, Missouri 65409, USA

MTA–DE Particle Physics Research Group, P.O. Box 51, H–4001 Debrecen, Hungary

MTA Atomki, P.O. Box 51, H–4001 Debrecen, Hungary

E-mail: ulj@mst.edu

Abstract. The proton radius puzzle is known as the discrepancy of the proton radius, obtained from muonic hydrogen spectroscopy (obtained as being roughly equal to 0.84 fm), and the proton radius obtained from (ordinary) hydrogen spectroscopy where a number of measurements involving highly excited states have traditionally favored a value of about 0.88 fm. Recently, a number of measurements of hydrogen transitions by the Munich (Garching) groups (notably, several hyperfine-resolved sublevels of the $2S$–$4P$) and by the group at the University of Toronto ($2S$–$2P_{1/2}$) have led to transition frequency data consistent with the smaller proton radius of about 0.84 fm. A recent measurement of the $2S$–$8D$ transition by a group at Colorado State University leads to a proton radius of about 0.86 fm, in between the two aforementioned results. The current situation points to a possible, purely experimental, resolution of the proton radius puzzle. However, a closer look at the situation reveals that the situation may be somewhat less clear, raising the question of whether or not the proton radius puzzle has been conclusively solved, and opening up interesting experimental possiblities at TRIUMF/ARIEL.

1. Introduction

We present a brief account of the current status of measurements of the proton radius, based on recent hydrogen spectroscopy [1–8] and scattering experiments [9, 10]. Also, we point out that a current discrepancy of the deuteron radius, based on deuterium spectroscopy and scattering experiments, merits further investigation [11]. Attention is drawn to the fact that in 1969, coindicidentally on the same day on which mankind set foot on the moon (21st of July, 1969), two Letters were published [12, 13], which report on an observation of a larger slope of the Sachs form factor of the proton, when measured using electron versus muon scattering, consistent with a (roughly) 2% difference in the proton radii derived from either scattering method. The current situation opens interesting experimental possibilities for the MUSE experiment [14–16] at the Paul–Scherrer Institute (PSI) and for a potential muon option at the Advanced Rare Isotope Laboratory (ARIEL) at TRIUMF (Vancouver).

2. Brookhaven Experiment

It might be useful to recall that scattering experiments on proton targets, using either electron or proton projectiles, have a considerable history. One example dates from about five decades ago [12, 13]. It was observed that muon-proton (μ-p) versus electron-proton (e-p) scattering data for the Sachs form factor taken in the range $0.15\,(\mathrm{GeV}/c)^2 < q^2 < 0.85\,(\mathrm{GeV}/c)^2$ differ from each other by a relative factor $N = 0.960 \pm 0.006$. It is stated, in the right text column on p. 154

of Ref. [12], that the "4% suppression of the form-factor ratio represents an 8% difference in the cross section since $\mathrm{d}\sigma/\mathrm{d}q^2 \propto G^2$".

One should note that the authors of Ref. [13] base their analysis on a number of assumptions. Let us consider the Rosenbluth formula [see Eq. (1) of Ref. [12]],

$$\left.\frac{\mathrm{d}\sigma}{\mathrm{d}q^2}\right|_{\mu,e} \left.\frac{\mathrm{d}\sigma}{\mathrm{d}q^2}\right|_{\mathrm{NS}} \frac{1}{\cot^2(\theta/2)} \left[2\tau G_M(q^2) + \frac{G_E^2(q^2) + \tau\, G_M(q^2)}{1+\tau} \cot^2(\theta/2)\right] , \tag{1}$$

where $\tau = -q^2/(4m_p^2) = \vec{q}^2/(4m_p^2)$ (for purely space-like momentum transfer), θ is the scattering angle, and $\mathrm{d}\sigma/\mathrm{d}q^2|_{\mathrm{NS}}$ is the differential cross section for spinless particles. The proton radius is inferred in Ref. [13] as follows [see also Eq. (11) of Ref. [17]],

$$r_p = \sqrt{\langle r^2 \rangle_p}\,, \qquad \langle r^2 \rangle_p = r_p^2 = 6\hbar^2 \left.\frac{\partial G_E(q^2)}{\partial q^2}\right|_{q^2=0} \tag{2}$$

The experiment [12, 13] has been carried out in a region of small scattering angle θ, where the second (slope) term in square brackets in Eq. (1) dominates the first (intercept), and the electric and magnetic form factors cannot be determined separately. The authors of of Refs. [13] base their analysis on the assumption that $G_E = G_M/\mu \equiv G$, where $\mu \approx 1$ is the ratio of the magnetic and electric form factors of the proton. This assumption is rather well justified and is in fact confirmed by more recent experiments [18].

In the abstract of Ref. [13], it is stated that the "apparent disagreement [of the μ-p versus e-p data] can possibly be accounted for by a combination of systematic normalization errors". This statement highlights a possible problem in the experiment [12, 13]: Namely, it is experimentally difficult to accurately normalize the flux of incoming particles, which could in principle account for the tentative muon-proton nonuniversality observed in Refs. [12,13]. Yet, as evident from the rather detailed analysis present in the two Letters [12, 13], it is evident that considerable efforts were made by the authors of Refs. [12,13] to avoid such normalization errors. In fact, in Ref. [12], the scattering data were checked for the possible presence of inelastic and multiple-scattering contributions, and also, the validity of the functional form (1) was verified against experimental data, for both muon as well as anti-muon projectiles. Let us also record an observation: The differential cross section is roughly proportional to the square of the Sachs electric G_E form factor (under the assumptions made above), and the slope of the Sachs form factor is again proportional to the square of the proton radius. So, the observed (roughly) 8% difference of the μ-p scattering cross sections [12, 13] translates into a 4% difference of the form factor, which translates into a 2% difference of the proton radii derived from μ-p versus e-p scattering. Tentatively, one could argue that, if the trend seen in the 1969 experiment were to be exclusively due to a normalization error, the relative systematic error would have to be on the order of 8% for the cross section, which is quite large.

Coincidentally, this is exactly the difference (both sign and magnitude agree) between the proton radii derived from the most recent $2S$–$8D$ measurement on atomic hydrogen [8] and the result from muonic hydrogen spectroscopy [4]. This situation gives a very pronounced motivation for the MuSE experiment [14–16], and also for adding a possible muon option to TRIUMF/ARIEL.

3. Brief Review of Hydrogen Experiments

In contrast to scattering experiments, the proton radius is derived from spectroscopic experiments via the finite-size energy shift, which, in leading order and applying the nonrelativistic approximation, reads as

$$\Delta E = \frac{2}{3} \frac{(Z\alpha)^4\,\mu c^2}{\pi\, n^3}\, \delta_{\ell 0} \left(\frac{\mu\, c\, r_p}{\hbar}\right)^2 , \tag{3}$$

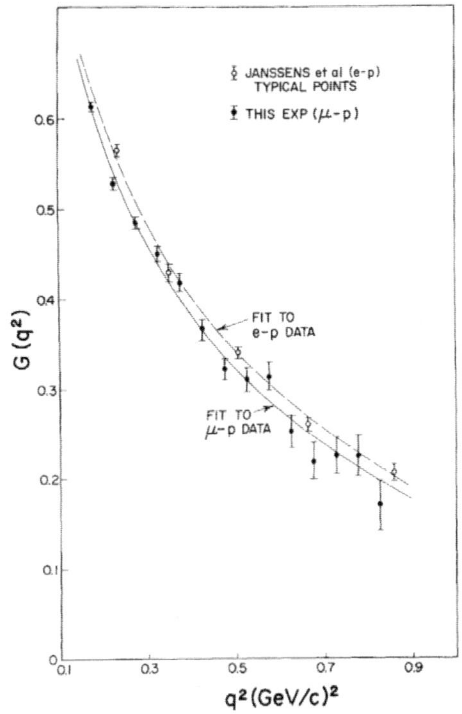

Figure 1. [Reproduced with permission.] Cross section data from Refs. [12, 13].

where Z is the nuclear charge number of the hydrogenlike ion, α is the fine-structure constant, μ is the reduced mass, c is the speed of light, and ℓ is the orbital angular momentum. The Kronecker δ implies that the energy correction is nonvanishing only for S states with $\ell = 0$. The relation (3) is only the leading approximation to the finite-size effect. Higher-order terms involve convoluted moments of the nuclear-charge distribution, as discussed in Chap. 5 of Ref. [19] and in great detail in Ref. [20].

Let us discuss Eq. (3). We have $Z = 1$ for both hydrogen and muonic hydrogen. The reduced mass μ is roughly 200 times larger for muonic bound systems as compared to ordinary hydrogen. The finite-size effect is proportional to μ^3, and thus, muonic hydrogen energy levels constitute a very sensitive probe of the proton radius.

In chronological order, let us discuss some recent results for the proton radius from spectroscopic [1–8], as well as scattering, experiments [9, 10]. We first recall that a least-squares fit of all experimental data available up to 2005 for both hydrogen as well as deuterium led to the results $r_p = 0.8750(68)$ fm ≈ 0.88 fm and $r_d = 2.1394(28)$ fm, based on a total of twenty-three measured transitions, including transitions to highly excited states [3].

In 2010, the muonic hydrogen $2S$–$2P_{1/2}$ Lamb shift measurement at PSI (Ref. [4]) led to a proton radius of $r_p = 0.84184(67)$ fm ≈ 0.84 fm. The result was in disagreement with the results of an electron scattering experiment carried out in Mainz (in 2010, see Ref. [9]), which led to the result of $r_p = 0.879(8)$ fm ≈ 0.88 fm, consistent with the value obtained from atomic hydrogen spectroscopys [3].

Since 2010, the discrepancy between the proton radius obtained from hydrogen spectroscopy ($r_p \approx 0.88$ fm) and from muonic hydrogen spectroscopy ($r_p \approx 0.84$ fm) has been termed the proton radius puzzle. In 2017, a measurement of the $2S$–$4P$ transition in atomic hydrogen (Ref. [6], taking into account various fine-structure and hyperfine sublevels) led to a result of

$r_p = 0.8335(95)$ fm, consistent with the smaller proton radius, otherwise obtained from muonic hydrogen spectroscopy. In the analysis of the experiment [6], emphasis is laid on so-called cross-damping terms, which had been analyzed (for differential cross sections) in Ref. [21].

A recent hydrogen measurement (in 2018, by the French group) of the $1S$–$3S$ transition [22] is consistent with the larger proton radius, with a result of $r_p = 0.877(13)$ fm ≈ 0.88 fm. In Ref. [5], it had been pointed out that the experimental approach taken by the Paris group should be largely independent of cross-damping terms.

The PRad Scattering Experiment, in 2019, obtained a value of $r_p = 0.831(14)$ fm, as reported in Ref. [10], based on scattering data. In 2019, a spectroscopic measurement in Toronto, of the $2S$–$2P_{1/2}$ Lamb shift in atomic hdrogen, resulted in a proton radius of $r_p = 0.833(10)$ fm [7], consistent with the smaller value of the proton radius. One might point that the $2P_{1/2}$–$2P_{3/2}$ fine-structure is nearly independent of the proton radius and can be calculated to very high precision [23]; its measurement would constitute an important consistency check for the smallness of the "Canadian protons" measured in Ref. [10]. (Of course, the Einstein equivalence principle (EEP) implies that the outcome of an experiment should be independent of where and when in the Universe it is performed. Assuming that protons are indistinguishable, identical elementary particles, one might be tempted to include the tongue-in-the-cheek remark that any conceivable size difference in "French versus Canadian and German Protons" would violate the EEP.)

In Ref. [5], it had been pointed out that the so-called cross-damping terms, which had played an important role in the analysis of the $2S$–$4P$ experiment [6], should be suppressed in $2S$–nD transitions in atomic hydrogen [see the discussion surrounding Eqs. (28)—(31) of Ref. [5]]. It had also been pointed out in Ref. [5] that cross-damping terms are "not likely to shift the experimental results reported in Refs. [1] and [2] on a level commensurate with the proton radius puzzle energy shift." We recall that Refs. [1] and [2] report on measurements of $2S$–nD transitions with $n = 8$ and $n = 12$. In consequence, it is of significance that a very recent (2022) measurement of the $2S_{1/2}$–$8D_{5/2}$ transition in atomic hydrogen (see Ref. [8]) has led to a proton radius of $r_p = 0.8584(51)$ fm ≈ 0.86 fm, which lies in between the previously accepted value of $r_p \approx 0.88$ fm from atomic hydrogen spectroscopy [3] and the value of $r_p \approx 0.84$ fm derived from muonic hydrogen spectroscopy [4].

The proton radius puzzle is exacerbated by the fact that the raw data for the Sachs form factor derived from the 2010 Mainz experiment [9] and the 2019 PRad data [10] are discrepant [24] (see also [25]). Clearly, further investigations are indicated.

4. Conclusions

In view of interesting possibilities with muon physics, one might ask if one should add a muonic beam to TRIUMF/ARIEL. One idea for low-cost muon sources can be mentioned, being based on an 8 GeV proton beam on a current-carrying target followed by a lithium lens and a quadrupole decay channel, where the decaying ions generate muons [26].

It is highly unlikely that muonic hydrogen theory, and Lamb shift theory, could provide explanations for the proton radius puzzle, since they are well under control. From the experimental side, the situation regarding the proton radius may be less clear than commonly thought. Electron versus muon scattering experiments could shed light. A confirmation of the 1969 experiment [12, 13], which could point to a small electron-muon nonuniversality, would be very desirable. A comparison of electron and muon scattering on the same apparatus would shed light on the issue [14]. Finally, let us remember that an even more striking 7σ disagreement persists for the deuteron radius, when comparing the results from electronic and muonic bound systems [11].

Acknowledgments
Insightful conversations with J. C. Bernauer are gratefully acknowledged. This research was supported by NSF grant PHY–2110294.

References

[1] B. de Beauvoir, F. Nez, L. Julien, B. Cagnac, F. Biraben, D. Touahri, L. Hilico, O. Acef, A. Clairon, and J. J. Zondy, Phys. Rev. Lett. **78**, 440 (1997).
[2] C. Schwob, L. Jozefowski, B. de Beauvoir, L. Hilico, F. Nez, L. Julien, F. Biraben, O. Acef, J. J. Zondy, and A. Clairon, Phys. Rev. Lett. **82**, 4960 (1999), [Erratum Phys. Rev. **86**, 4193 (2001)].
[3] U. D. Jentschura, S. Kotochigova, E.-O. Le Bigot, P. J. Mohr, and B. N. Taylor, Phys. Rev. Lett. **95**, 163003 (2005).
[4] R. Pohl *et al.*, Nature (London) **466**, 213 (2010).
[5] U. D. Jentschura, Phys. Rev. A **92**, 012123 (2015).
[6] A. Beyer, L. Maisenbacher, A. Matveev, R. Pohl, K. Khabarova, A. Grinin, T. Lamour, D. C. Yosta, T. W. Hänsch, N. Kolachevsky, and T. Udem, Science **358**, 79 (2017).
[7] N. Bezginov, T. Valdez, M. Horbatsch, A. Marsman, A. C. Vutha, and E. A. Hessels, Science **365**, 1007 (2019).
[8] A. D. Brandt, S. F. Cooper, C. Rasor, Z. Burkley, A. Matveev, and D. C. Yost, Phys. Rev. Lett. **128**, 023001 (2022).
[9] J. C. Bernauer, P. Achenbach, C. Ayerbe Gayoso, R. Böhm, D. Bosnar, L. Debenjak, M. O. Distler, L. Doria, A. Esser, H. Fonvieille, J. M. Friedrich, J. Friedrich, M. Gómez Rodríguez de la Paz, M. Makek, H. Merkel, D. G. Middleton, U. Müller, L. Nungesser, J. Pochodzalla, M. Potokar, S. Sánchez Majos, B. S. Schlimme, S. Sirca, T. Walcher, and M. Weinriefer, Phys. Rev. Lett. **105**, 242001 (2010).
[10] W. Xiong *et al.*, Nature (London) **575**, 147–150 (2019).
[11] R. Pohl *et al.*, Science **353**, 669 (2016).
[12] L. Camilleri, J. H. Christenson, M. Kramer, L. M. Lederman, Y. Nagashima, and T. Yamanouchi, Phys. Rev. Lett. **23**, 149 (1969).
[13] L. Camilleri, J. H. Christenson, M. Kramer, L. M. Lederman, Y. Nagashima, and T. Yamanouchi, Phys. Rev. Lett. **23**, 153 (1969).
[14] see https://www.psi.ch/en/muse/experiment.
[15] M. Kohl *et al.* [MUSE Collaboration], Eur. Phys. J. Web of Conferences **66**, 06010 (2014).
[16] M. Kohl, private communication (2022).
[17] U. D. Jentschura, Eur. Phys. J. D **61**, 7 (2011).
[18] M. K. Jones *et al.*, Phys. Rev. Lett. **84**, 1398 (2000).
[19] U. D. Jentschura and G. S. Adkins, *Quantum Electrodynamics: Atoms, Lasers and Gravity* (World Scientific, Singapore, 2022).
[20] J. L. Friar, Ann. Phys. (N.Y.) **122**, 151 (1979).
[21] U. D. Jentschura and P. J. Mohr, Can. J. Phys. **80**, 633 (2002).
[22] H. Fleurbaey, S. Galtier, S. Thomas, M. Bonnaud, L. Julien, F. Biraben, F. Nez, M. Abgrall, and J. Guéna, Phys. Rev. Lett. **120**, 183001 (2018).
[23] U. Jentschura and K. Pachucki, Phys. Rev. A **54**, 1853 (1996).
[24] J. C. Bernauer, private communication (2022).
[25] U. D. Jentschura, Talk given at the workshop "New Scientific Opportunities and the TRIUMF ARIEL e-linac", TRIUMF, Vancouver, Canada, 26th of May, 2022.
[26] V. Balbekov and N. Mokhov, *Low-Budget Muon Source*, PAC2001, TPAH144, Particle Accelerator Conference, FNAL, Batavia, Illinois, 2001.

Electrodisintegration of ^{16}O and determination of astrophysical S-factors of the inverse reaction

I. Friščić[1] , T.W. Donnelly[2] and R.G. Milner[2]

[1] Department of Physics, Faculty of Science, University of Zagreb, Bijenička c. 32, 10000 Zagreb, Croatia

[2]Laboratory for Nuclear Science, Massachusetts Institute of Technology, Cambridge, MA 02139, USA

E-mail: ifriscic@phy.hr

Abstract.
We performed Monte Carlo calculations in a simple model that relates the radiative capture and electrodisintegration reactions and showed that by using a windowless gas jet target and modern energy-recovery linear accelerators (ERLs, CBETA at Cornell, MESA at Mainz, ARIEL e-linac at TRIUMF) to reach high luminosity, a high precision measurement of the electrodisintegration of the ^{16}O nucleus would provide a method to determine the rate of the α on ^{12}C radiative capture with a superb statistical precision compared to existing data. The calculations were done for the beam energies of 30, 50 and 114 MeV and show a significant reduction in terms of statistical uncertainties for the energy range $E_{c.m.} < 2$ MeV. The 114 MeV setting has the smallest statistical uncertainty, but at the same time the calculations for this beam energy setting were performed the furthest away from the real photon limit. The feasibility of this simple model still needs to be studied experimentally. This could be done with an experiment focused at $E_{c.m.} > 2$ MeV region, where the cross section is much larger.

1. Introduction

After more than half a century of experimental effort, the rate of the $^{12}\mathrm{C}(\alpha,\gamma)^{16}\mathrm{O}$ reaction – one of the key nuclear inputs in every stellar evolution model – is still not known with a desired accuracy at the the astrophysically relevant energy. Even for variations in the value of the reaction rate in the range of few standard deviations and the fixed initial mass of the massive star, the stellar evolution models produce dramatically different results in terms of the $^{12}\mathrm{C}/^{16}\mathrm{O}$ ratio at the end of helium burning stage, abundance of the elements in the subsequent nucleosynthesis and remnants at the end of the stellar lifetime [1–5].

There have been different approaches to improve the accuracy of the $^{12}\mathrm{C}(\alpha,\gamma)^{16}\mathrm{O}$ reaction rate ranging from astrophysical measurements, such as: observation of white dwarfs [6], galactic chemical evolution model [7] and gravitational-wave detection of binary black hole mergers [4], to terrestrial experiments which have in common the strategy to measure the rate at some larger energies and then extrapolate to the astrophysical relevant energy – Gamow energy ($\sim$ 300 keV). These experiments are summarized in [8] and include measurements of direct reaction [9–22], elastic scattering $^{12}\mathrm{C}(\alpha,\alpha)^{12}\mathrm{C}$ [23, 24] and $\beta-$delayed α-decay of ^{16}N [25–27].

Recently, two approaches based on photodisintegration of ^{16}O by a real photon beam $q = \omega$ (q is the 3-momentum and ω is the energy of the photon) have started to be validated, including a bubble chamber experiment [28, 29] and an optical time projection chamber experiment [30, 31]. The disintegration of the ^{16}O nucleus can also be induced by a virtual photon $q > \omega$ coming from electron scattering [32]. The recent interest in measuring the electrodisintegration of ^{16}O [33–37] was motivated by development of new jet gas targets [38] and new generation of energy-recovery linear (ERL) electron accelerators like CBETA [39], MESA [40] and ARIEL e-linac at TRIUMF [41, 42], which would allow for large luminosity $>10^{35}$ cm^{-2} s^{-1}.

In electrodisintegration, the energy ω and the 3-momentum q of the virtual photon, which are transferred to the final state α-particles, can be controlled separately by selecting the electron beam energy E_e, angle of the scattered electron θ_e and the energy of the scattered electron E'_e. Therefore, at fixed ω it is possible to increase the kinetic energy of the produced α-particle by increasing the transferred 3-momentum q, thus ensuring that the α-particle exits the jet target with some moderate angular spread [33].

At low energies, only the E1 (electric dipole) and E2 (electric quadrupole) multipole contribute to the cross section of reactions involving real photons [9]. The contributions of each multipole can be separated either by measuring the angular distribution of the produced gammas in case of the direct reaction [9] or by measuring the the angular distribution of the produced α-particles for inverse reaction [31]. In an exclusive electrodisintegration experiment, the final states of the scattered electron and the α-particle need to be measured in coincidence. The final state of the unobserved ^{12}C can be determined from the kinematics and any excited state can be excluded. The multipoles can be separated by using the angular distribution of the measured α-particles.

In this paper, we go through the key steps in building a simple model and a formalism which relates the ^{12}C$(\alpha,\gamma)^{16}$O reaction and the electrodisintegration of ^{16}O. A detailed description can be found in [33]. The model is then used to calculate the rate of electrodisintegration of ^{16}O for electron beam energies of 30, 50 and 114 MeV, and subsequently to estimate the statistical uncertainties of S_{E1}- and S_{E2}-factors. At the end, we briefly discuss the simplicity of the model relating to extrapolation to the real photon limit, possible problems and solutions.

2. Model and Monte Carlo Calculations

For the unpolarized ^{16}O target and the unpolarized electron beam, the differential cross section for the electrodisintegration of ^{16}O in the *c.m.* frame can be written as:

$$\left[\frac{d\sigma}{d\omega d\Omega_e d\Omega_\alpha^{c.m.}}\right]_{(e,e'\alpha)} = \frac{M_\alpha M_{^{12}C}}{8\pi^3 W}\frac{p_\alpha^{c.m.}\sigma_{Mott}}{(\hbar c)^3}\left(\widetilde{v}_L R_L + \widetilde{v}_T R_T + \widetilde{v}_{TL} R_{TL} + \widetilde{v}_{TT} R_{TT}\right). \quad (1)$$

where M_α and $M_{^{12}C}$ are the masses of the produced α-particle and ^{12}C nucleus, $p_\alpha^{c.m.}$ is the 3-momentum of the α-particle in *c.m.* frame, $W = \sqrt{(M_{^{16}O}+\omega)^2 - q^2}$ is the total invariant mass of the virtual photon + ^{16}O system, σ_{Mott} is the Mott cross section, functions $\widetilde{v}_K$ are kinematical factors of the electron in the *c.m.* frame and the R_K are response functions. The cross section formulas describing the reactions with real photons contain only the transverse response function R_T for unpolarized photons and the transverse interference response R_{TT} if there are linearly polarized photons involved [33]. In the real photon limit ($\omega = q$), R_T is the same function for all three types of reactions (direct, photodisintegration and electrodisintegration).

In the multipole decomposition of response functions, the longitudinal response R_L contains only Coulomb matrix elements t_{CJ}, the transverse R_T and the transverse interference response R_{TT} only electric matrix elements t_{EJ} and the transverse-longitudinal response function R_{TL} the mixture of both matrix elements. In the case of q being small compared to the typical nuclear scale for the 3-momentum denoted as $q_0 \sim 230$ MeV/c, t_{CJ} and t_{EJ} may be parameterized as:

$$t_{CJ}(q) \equiv \left(\frac{q}{q_0}\right)^J a'_{CJ}\left[1 + \left(\frac{q}{q_0}\right)^2 b'_{CJ}(q)\right] e^{-(q/q_0)^2} \quad (2)$$

$$t_{EJ}(q) \equiv \left(\frac{\omega}{q}\right)\left(\frac{q}{q_0}\right)^J a'_{EJ}\left[1+\left(\frac{q}{q_0}\right)^2 b'_{EJ}(q)\right] e^{-(q/q_0)^2} \tag{3}$$

By using the long wavelength limit ($q \ll q_0$) and the continuity equation we can write:

$$\lim_{q \ll q_0} \sqrt{\frac{J}{J+1}} t_{EJ}(q) = -\left(\frac{\omega}{q}\right) t_{CJ}(q), \quad a'_{EJ} = -\sqrt{\frac{J+1}{J}} a'_{CJ} \quad for \quad J \geq 1. \tag{4}$$

By taking the real photon limit for the electric matrix element and inserting it into the differential cross section formula for the direct reaction it is possible to obtain a simple relation between leading-order coefficients a'_{EJ} and the corresponding astrophysical S_{EJ}-factor. Therefore, the leading-order coefficients a'_{EJ} were determined by taking all experimental S_{EJ}-factors below $E_{c.m.} < 1.7$ MeV into account and fitting a second-order polynomial to model the energy dependence of the S_{EJ}-factor in the energy region 0.7 MeV$< E_{c.m.} < 1.7$ MeV, as shown in Figure 1.

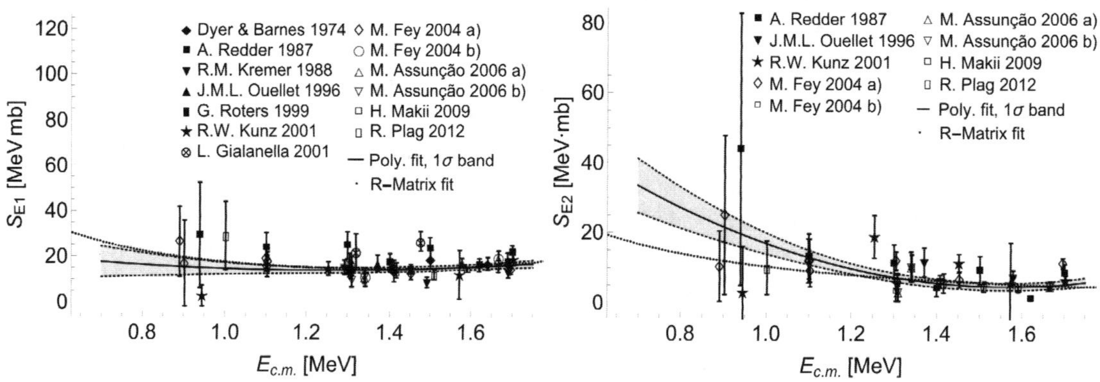

Figure 1. Measured S_{E1}- and S_{E2}-factors from previous experiments [9–17, 19, 20, 22]. The dotted line represents the AZURE2 [43] R-Matrix fit of the world data set and the solid line is second-order polynomial fit of the experimental S_{EJ}-factor data in the energy range 0.7 MeV$< E_{c.m.} < 1.7$ MeV.

We assumed ideal detector performance, a magnetic spectrometer for the electrons at a scattering angle of $\theta_e = 15°$ and α-particle detectors covering the azimuth angle range $\pi/2 \leq \phi_\alpha^{c.m.} \leq 3\pi/2$ and $0° \leq \theta_\alpha^{c.m.} \leq 60°$ range for the polar angle. The top-view layout of the setup is shown in Figure 2. Compared to the calculations reported in [33–35] where full $0 \leq \phi_\alpha^{c.m.} \leq 2\pi$ is used, here only α-particles which are generated left with respect to the direction of the virtual photon are used in the calculations. This cut was chosen to exclude the fraction of α-particles generated on the right side of the virtual photon direction, which would in the cases of the two lowest beam energy settings overlap with the beam pipe and the electron spectrometer (angle θ_{γ^*} in Figure 2 is not large enough to ensure full $0° \leq \theta_\alpha^{c.m.} \leq 60°$ coverage).

At the moment, there are no experimental data describing the behavior of the next-to-leading order (NLO) coefficients b'_{EJ}, b'_{CJ} and the Coulomb multipole $C0$. Therefore, the NLOs were approximated to have the same order of magnitude as the associated leading-order coefficient a' and assumed to have a positive sign, *i.e* $b'_{EJ} \approx +1$ and $b'_{CJ} \approx +1$, and the Coulomb multipole $C0$ was approximated with a'_{E2} (so called case A from [33]). For the phase shifts which also enter into response functions, only Coulomb phase shifts have been taken into account, since for

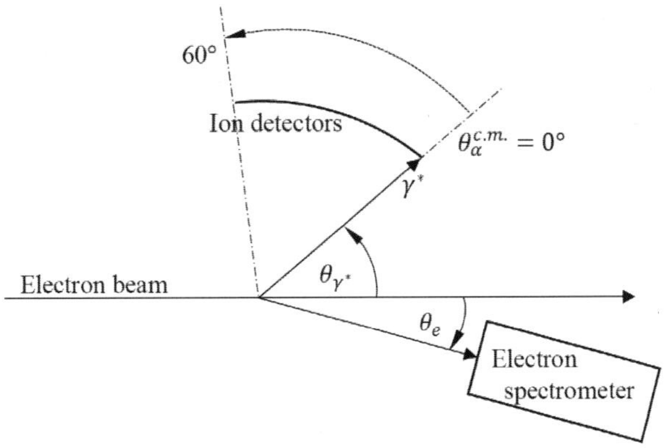

Figure 2. Top-view layout of the setup defining limits used in the Monte Carlo calculations.

energies $E_{c.m.} < 1.7$ MeV elastic phase shifts of the s, p and d waves are almost zero [23, 24]. 100 days of data taking was assumed with an electron beam of intensity 10 mA and an oxygen jet target capable of an areal thickness of 5×10^{18} atoms/cm^2, which amounts to a luminosity of 3.125×10^{35} cm^{-2}s^{-1}. Monte Carlo calculations of the number of events collected were performed for the three different electron beam energies 30, 50 and 114 MeV. The ARIEL e-linac is designed to produce a 30 MeV beam at 10 mA and in the near future the beam energy will be upgraded to 50 MeV [41, 42]. The beam of 114 MeV corresponds to CBETA [39] and similar calculations have been already reported in [33] but for a beam current of 40 mA.

3. Results and Discussion

The results of the Monte Carlo calculations for the electron beam energies of 30, 50 and 114 MeV in terms of the projected relative statistical uncertainties are shown in Figure 3. When compared to the relative uncertainties of the experimental data, the calculations for electrodisintegration of ^{16}O show a clear advantage in terms of statistical uncertainties.

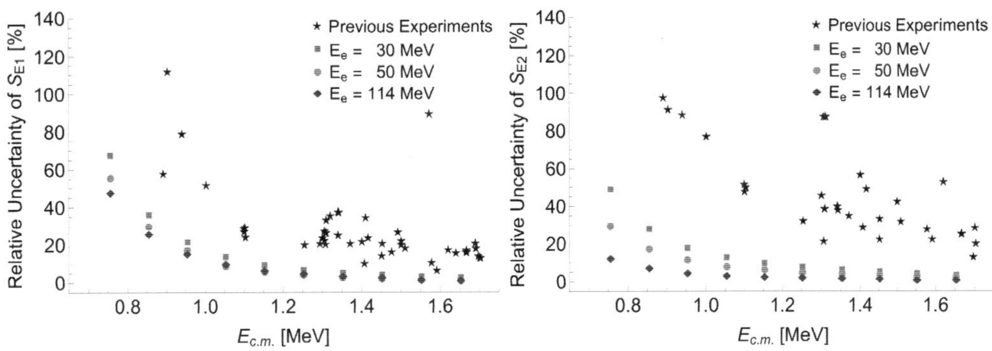

Figure 3. Relative statistical uncertainties of the S_{E1}- and S_{E2}-factors for beam energies $E_e = 30$, 50 and 114 MeV. Solid stars represent relative uncertainties of S_{E1}- and S_{E2}-factors determined in previous experiments [9–17, 19, 20, 22].

The results in Figure 3 suggest that for fixed electron scattering angle θ_e and fixed measuring time one would obtain better statistical precision for a larger beam energy, but at the cost of being further away from the real photon limit ($q_{E=30MeV} < q_{E=50MeV} < q_{E=114MeV}$). By choosing the parametrization of the Coulomb and electric matrix elements, as defined in equations 2 and 3, extracting the leading-order coefficients with corresponding statistical uncertainties is reduced to the problem of fitting and subsequent straight-forward conversion to astrophysical S_{EJ}-factors with associated statistical uncertainties. In general, the problem of extrapolation to the real photon limit may not be so easy to solve. Things would get more complicated very fast, if the leading-order coefficients a'_{EJ} would also be functions of transferred 3-momentum q, or if the expansion of the matrix elements up to the NLO may not converge properly and next-to-next-to-leading would be required.

If such situations occur, the solution would be to perform the measurement of the ^{16}O electrodisintegration as close as possible to the real photon limit. Either by using a lower electron beam energy and suddenly making the ARIEL e-linac's lower energy a more favourable option or by setting the electron scattering angle to zero, as discussed in the framework of MAGIX collaboration for the"phase 2" measurement with a zero degree tagger [37].

4. Conclusion and Outlook

By using a simple model which relates the radiative capture and electrodisintegration reactions [33], we performed Monte Carlo calculations and projected statistical uncertainties for the S_{E1}- and S_{E2}-factors assuming the selected beam parameters available at the modern ERLs (10 mA beam current, beam energies of 30, 50 and 114 MeV). For all three beam energy settings, the calculated statistical uncertainties are a substantial improvement over uncertainties of the existing experimental data and the setting with 114 MeV beam energy, also being furthest from the real photon limit, turned out to be the better choice compared to the other two beam energies.

The simple model [33] was constructed in a way to make the extrapolation of the electrodisintegration calculations or measurements to the real photon limit simple. An initial, shorter experiment focusing on the energy range $E_{c.m.} > 2$ MeV where the cross section is larger, thus not requiring the extreme beam current, would be suitable to experimentally study the extrapolation to the real photon limit. Furthermore, such an experiment would be also helpful to verify other assumptions and approximations which were not the focus of this paper, but are also important [33], as well as to study the problem of operating the α-detectors in close proximity to the Megawatt electron beam and systematic uncertainties related to the specific experimental setup.

Acknowledgments

I. Friščić is currently supported by Croatian Science Foundation under the project IP-2018-01-8570 and European Union's Horizon 2020 research and innovation program under the grant agreement 101038099.

References

[1] Weaver T A and Woosley S E 1993 *Phys. Rep.* **227** 65–96
[2] Woosley S E, Heger A and Weaver T A 2002 *Rev. Mod. Phys.* **74** 1015
[3] Woosley S E *et al.* 2003 *Nucl. Phys. A* **718** 3c
[4] Farmer R *et al.* 2020 *Astrophys. J. Lett.* **902** L36
[5] Aliotta M *et al.* 2022 *J. Phys. G: Nucl. Part. Phys.* **49** 010501
[6] Metcalfe T S 2003 *Astrophys. J.* **587** L43
[7] Woosley S E, Timmes F X and Weaver T A 1993 *J. Phys. G: Nucl. Part. Phys.* **19** S183
[8] deBoer R J *et al.* 2017 *Rev. Mod. Phys.* **89** 035007

[9] Dyer P and Barnes C 1974 *Nucl. Phys. A* **233** 495
[10] Redder A *et al.* 1987 *Nucl. Phys. A* **462** 385
[11] Kremer R M *et al.* 1988 *Phys. Rev. Lett.* **60** 1475
[12] Ouellet J M L *et al.* 1996 *Phys. Rev. C* **54** 1982
[13] Roters G, Rolfs C, Strieder F and Trautvetter H 1999 *Eur. Phys. J. A* **6** 451
[14] Gialanella L *et al.* 2001 *Eur. Phys. J. A* **11** 357
[15] Kunz R *et al.* 2001 *Phys. Rev. Lett.* **86** 3244
[16] Kunz R W 2002 Ph.D. thesis Univ. of Stuttgart
[17] Fey M 2004 Ph.D. thesis Univ. of Stuttgart
[18] Schürmann D *et al.* 2005 *Eur. Phys. J. A* **26** 301–305
[19] Assunção M *et al.* 2006 *Phys. Rev. C* **73** 055801
[20] Makii H *et al.* 2009 *Phys. Rev. C* **80** 065802
[21] Schürmann D *et al.* 2011 *Phys. Lett. B* **703** 557–561
[22] Plag R *et al.* 2012 *Phys. Rev. C* **86** 015805
[23] Plaga R *et al.* 1987 *Nucl. Phys. A* **465** 291–316
[24] Tischhauser P *et al.* 2009 *Phys. Rev. C* **79** 055803
[25] Buchmann L *et al.* 1993 *Phys. Rev. Lett.* **70** 726
[26] Azuma R E *et al.* 1994 *Phys. Rev. C* **50** 1194
[27] Tang X D *et al.* 2010 *Phys. Rev.C* **81** 045809
[28] DiGiovine B *et al.* 2015 *Nucl. Instrum. Methods Phys. Res. A* **781** 96–104
[29] Holt R J *et al.* 2018 (*Preprint* 1809.10176)
[30] Gai M *et al.* 2010 *JINST* **5** P12004
[31] Smith R, Gai M, Stern S R, Schweitzer D K and Ahmed M W 2021 *Nat. Commun.* **12** 5920
[32] Raskin A S and Donnelly T W 1989 *Ann. Phys.* **191** 78–142
[33] Friščić I, Donnelly T W and Milner R G 2019 *Phys. Rev. C* **100** 025804
[34] Friščić I, Donnelly T W and Milner R G 2020 *Proc. of the 63rd ICFA Advanced Beam Dynamics Workshop on Energy Recovery Linacs (ERL19)* 18–21
[35] Friščić I, Donnelly T W and Milner R G 2020 *J. Phys. Conf. Ser.* **1643** 012056
[36] Holt R J and Filippone B W 2019 *Phys. Rev. C* **100** 065802
[37] Lunkenheimer S 2022 *Feasibility Studies for Measuring the Astrophysical S-Factor of the Reaction* $^{12}C(\alpha,\gamma)^{16}O$ *via Electro-Disintegration at MAGIX* Ph.D. thesis Johannes Gutenberg, Univ. of Mainz
[38] Grieser S *et al.* 2018 *Nucl. Instrum. Methods Phys. Res. A* **906** 120–126
[39] Hoffstaetter G H *et al.* 2017 (*Preprint* 1706.04245)
[40] Hug F *et al.* 2017 *Proc. of Linear Accelerator Conference (LINAC'16)* **28** 313–315
[41] Laxdal R E, Ang Z, Au T, Fong K *et al.* 2018 *Proc. of 18th International Conference on RF Superconductivity (SRF17)* 6–12
[42] Ma Y, Ang Z, Fong K, Keir J J *et al.* 2018 *Proc. of the 9th International Particle Accelerator Conference* 4512–4515
[43] Azuma R E *et al.* 2010 *Phys. Rev. C.* **81** 045805

Journal of Physics: Conference Series **2391** (2022) 012019 doi:10.1088/1742-6596/2391/1/012019

Closing Remarks

Richard G. Milner

Massachusetts Institute of Technology, Cambridge, MA, USA

E-mail: milner@mit.edu

Abstract. This paper summarizes the workshop and presents a perspective on the path forward.

1. Summary

Over the last two and a half days, we have heard twenty-three stimulating talks on scientific opportunities, ongoing and planned activities, and accelerator developments from around the world pertinent to the ARIEL electron-linac. It is truly an exciting time for low-energy (< 300 MeV) electron scattering experiments to pursue studies in subatomic physics: storage rings (SCRIT/RIKEN), linacs (S-DALINAC/Darmstadt), special secondary beams (MUSE/PSI), ERLs (MESA/Mainz) and others. While the search for new-physics beyond the Standard Model (SM) at low energies has been a principal driver for this workshop, there are also clear scientific opportunities to continue to study the physical world within the SM. Examples include the electromagnetic structure of the nucleon and few-body nuclei at low momentum transfers as well as the study of nuclear structure, particularly for short-lived nuclei.

It is striking that in North America, while there have been major contributions to novel, low-energy, high-intensity electron-beam development, e.g. at LERF/FEL/JLab and CBETA/Cornell, there is no planned subatomic physics program in low-energy electron scattering. Thus, the ARIEL electron linac offers an unique opportunity for N. America to pursue studies in subatomic physics with low-energy electron scattering. The scientific opportunities are well summarized in many of the contributions to this proceedings.

As we embark on searches for new physics, it is instructive to remember that in 1970 the best experiments had excluded new physics around the dilepton invariant mass of 3.1 GeV. See Fig. 1 which shows data from [1] that excluded new physics at 3.1 GeV well below the rate for production of the J/ψ, which was discovered [2] in 1974, as shown in Fig. 2. Note that both experiments were done with a proton beam incident on nuclear targets at the same accelerator, the Alternating Gradient Synchrotron (AGS) at Brookhaven National Laboratory. How can this be? Basically the 1974 experiment had a much better energy resolution for the mass of the dimuon pair. Indeed, in Fig. 1 one can clearly see a shoulder on the data starting at 3 GeV. However, the mass energy resolution was not sufficient to see the peak discovered in 1974. The lesson here is that though the parameter space for new physics may be excluded at 2σ, a new, different experiment with improved characteristics to explore the same region is worthwhile and may lead to a discovery. Logically, exclusion limits cannot prove that new physics does not exist. Conversely, any hints of evidence for new physics must be pursued relentlessly.

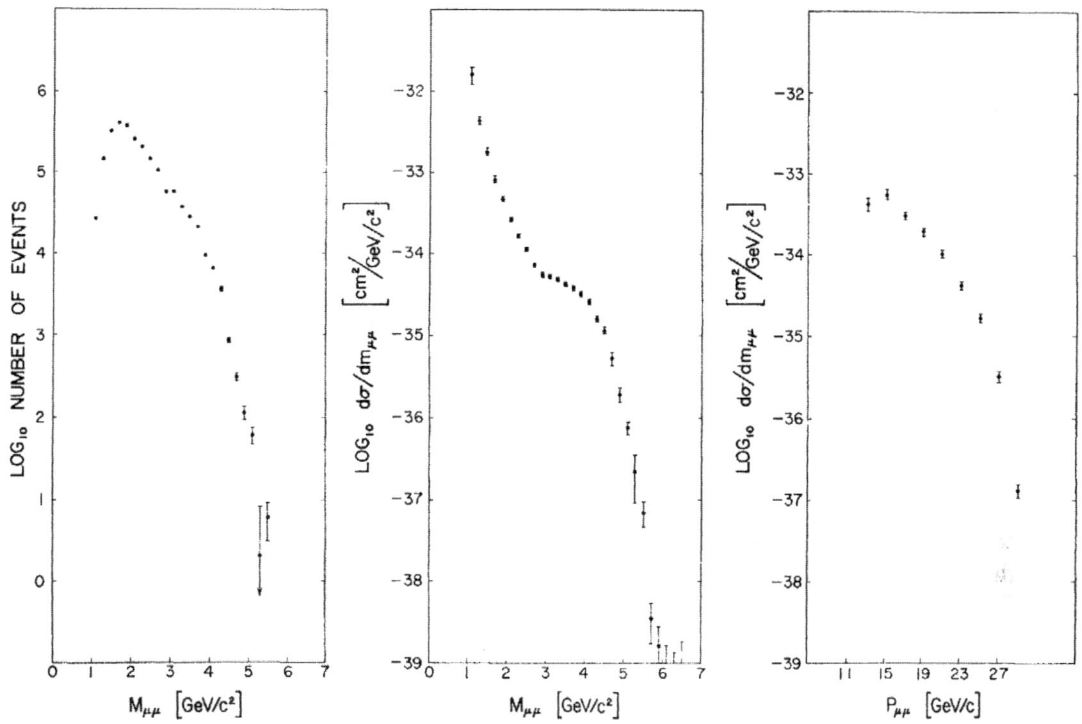

Figure 1. From [1]: (a) Observed events as a function of the effective mass of the muon pair. (b) Cross section as a function of the effective mass of the muon pair (these data include the wide-angle counters). (c) Cross section as a function of the laboratory momentum of the muon pair.

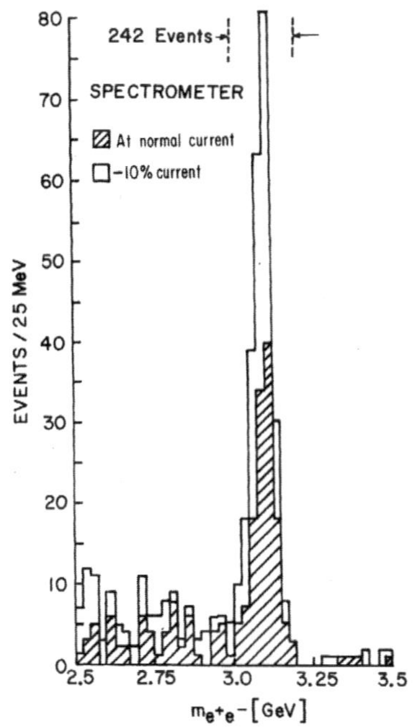

Figure 2. From [2]: mass spectrum showing the existence of J. Results from two spectrometer settings are plotted showing that the peak is independent of spectrometer currents. The run at reduced current was taken two months later than the normal run.

2. The Path Forward

The experiment DarkLight@ARIEL [3] must continue to make progress. This involves: developing the necessary electron beam with characteristics in terms of halo, background, stability, intensity etc.; finalizing the design of the spectrometer magnets and target at the optimal location; securing the necessary funding to construct the experiment; constructing the experiment; installing and commissioning the experiment; and taking data.

In terms of the broader scientific opportunities with the ARIEL electron linac, there has been really excellent discussion at this workshop. This is summarized in the opening contribution to this proceedings. This serves as an excellent starting point for further consideration and for discussion at the long range planning in nuclear and particle physics underway in Canada, Europe and the U.S. at present.

I am grateful for valuable discussions with my workshop organizing committee colleagues as well as with Susan Gardner and Toshimi Suda. The author's research is supported by the US Department of Energy Office of Nuclear Physics.

References

[1] J.H. Christenson et al. Observation of massive muon pairs in hadron collisions. *Phys. Rev. Lett.*, 25:1523, November 1970.
[2] J.J. Aubert et al. Experimental observation of a heavy particle J. *Phys. Rev. Lett.*, 33:1404, December 1974.
[3] E. Cline and others (The DarkLight Collaboration). Searching for New Physics with DarkLight at the ARIEL Electron-Linac. *arXiv:2208.1402*, August 2022.

AUTHOR INDEX